C000319156

Searching the Internet

Gilles Fouchard

**With additional material by
Rob Young**

An imprint of PEARSON EDUCATION

PEARSON EDUCATION LIMITED

Head Office:
Edinburgh Gate
Harlow
Essex CM20 2JE
Tel: +44 (0)1279 623623
Fax: +44 (0)1279 431059

London Office:
128 Long Acre
London WC2E 9AN
Tel: +44 (0)20 7447 2000
Fax: +44 (0)20 7240 5771

First published in Great Britain 1999

First published in 1999 as
Recherche sur Internet – Se Former en 1 Jour
by CampusPress France
19, rue Michel Le Comte
75003 Paris
France

Library of Congress Cataloging in Publication Data
Available from the publisher.

British Library Cataloguing in Publication Data
A CIP catalogue record for this book can be obtained from the British Library.

ISBN 0-13-021249-0

10 9 8 7 6 5 4 3 2

Translated and typeset by Berlitz Translation Services UK, Baldock, Hertfordshire.
Printed and bound in Great Britain by Redwood Books Ltd, Trowbridge.

The publishers' policy is to use paper manufactured from sustainable forests.

Table of contents

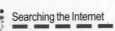

Introduction

The Internet in general and the Web in particular make up a veritable labyrinth. After a few hours of solo surfing, the surfer may feel truly frustrated. The overabundance of information and random access to every kind of information cannot be the real purpose of the Internet. It is the practical use that we make of the Internet which creates the need for specific searches. To obtain the latest updates (on anything from news to music releases), prepare a report on radioactivity, look at the works of a famous artist, download software, listen to foreign radio, rent a cottage or even buy a rare book, you need to use the Internet's search tools. For this, the Web offers us hundreds of services, search engines and help tools to make our investigations on the Internet easier. But we still need to be able to identify the right tool and to learn to use it efficiently. That is the object of this book.

Each chapter will only take one hour to learn. From acquiring basic skills to using the most advanced search engines, including specialist search techniques, every aspect of browsing the Internet will be opened up to you. In one day you can become an accomplished surfer on the Internet.

PURPOSE OF THIS BOOK

Searching the Internet is both:

- a **complete guide** to looking for information, people, services and objects of every kind (images, software, etc.) on the Internet;

- a **teaching method** presenting in twelve chapters (one hour per chapter) all the techniques and the best search services available on the Web.

This pocket guide is aimed at:

- **Beginners**. They will thus be able to take their first steps on the Web, use fast search methods and discover the search resources needed to locate UK sites or people.

- **Experienced surfers**. They will use the metasearch engines, discover advanced search techniques and multilingual browsing and learn how to get the most out of search engines.

- **Professionals**. They will be able to search documents and use professional directories and services.

It will be indispensable for:

- getting to grips with the vast range of information on the Web;

- getting to know search tools;

- learning to use search engines;

- learning to make the right choice of tool;

- mastering advanced search techniques;

- using "push" technologies properly.

The icons in the margin indicate notes providing additional information, explaining a new idea or presenting a term used for the first time. The icon used tells you about the content of the note.

These notes provide additional information about the subject concerned.

These notes indicate a variety of shortcuts: keyboard shortcuts, "wizard" options, techniques reserved for experts, etc.

These notes warn you of the risks associated with a particular action and, where necessary, show you how to avoid any pitfalls.

ABOUT THE AUTHOR

An engineer by training (Lille Ecole Centrale) Gilles Fouchard is by turns software developer, Oracle consultant (database), author and magazine designer.

Since 1992 he has devoted himself to the launch and development of magazines. *Multimédia Solutions* (Edicorp), the leading French magazine devoted to multimedia technology, then *CD-ROM Magazine, Home PC, Internet Guide du Web* (Sepcom) and *Pcmag Loisirs*.

Since the end of 1997, Gilles Fouchard (**fouchard@planetepc.fr**) has been an Internet consultant (on-line databases, trade and magazines) for various companies.

Chapter 1

Basic skills

THE CONTENTS FOR THIS CHAPTER

- What is a search engine?
- What does the search engine consist of?
- Searching the Internet: principles
- Getting started with your navigator

People usually measure the Internet and its growth in terms of the number of users. With around 50 million surfers throughout the world, and around 1 million new users every month, the Internet is a phenomenon which cannot be ignored. Its content (number of sites, number of Web pages) is also growing at a phenomenal rate. In July 1997 there were 20 million servers (host) compared with 13 million just one year earlier. In the same way, the increase in domain names is significant of the number of sites set up on the Web: 1.3 million in July 1997 compared with just 500,000 one year earlier. You can imagine the effect of this on the number of

Web pages! With such exponential growth, the Web is becoming a veritable labyrinth through which it is difficult to find one's way. Whether a novice or an experienced surfer, you need to use search tools in order to make effective use of the Internet.

To all this we need to add the growth of *newsgroups*, downloading sites (ftp) and Web channels. The latter are the result of *push* (as opposed to *pull*) technologies. With a Web channel, information comes to you, whereas most of the time the Web obliges you to "pull" information using search engines. But you still need to know how to find the right channels and to use them proficiently.

In order to achieve what you set out to do, avoiding time wasted surfing aimlessly, it is necessary to use the best tools offered on the Internet in general and the Web in particular. At your service you have search engines, directories, help features, selections, metasearch engines and "searchbots". As you will see in this book, there are plenty of good tools. So you will need to make the right choice and then you will need to learn how to use these tools. Amongst this plethora of tools, the search engine is of unimagined power for the beginner. Experienced surfers will also be able to refine their technique and learn how to select the right tools, optimise searches and interpret results.

Searching the Internet is empirical in nature. You can either work strictly in order to achieve your goal without any detours, or you can proceed by means of trial and error in order to uncover the odd pearl. It is by searching that you learn to search, so don't hesitate to try things out, to overlap requests and to refine your methods and how you express yourself as you progress.

WHAT IS A SEARCH ENGINE?

For anybody who has already ventured onto the Web, this question does not arise. The search tool on the Internet is as indispensable as a library's catalogue, or the index of an encyclopaedia.

On the Web, the use of an index is that it can label millions of pages and even billions of words. Like any document, a Web page may be indexed in two different ways:

- using keywords defining the page or document;
- using full text mode indexing.

In the first case, you can, for example, find all the pages relating to "Tourism" (it is the page author who has indicated the keyword, or a professional surfer who has labelled this page in this category). In the second case, you will have access to all pages containing the word "tourism".

A search engine therefore indexes Internet objects (sites, Web pages, objects contained on a page, etc.). This strength is doubled by the updating of these indexes. Hundreds of sites are being created around the world every day and in line with this the content of indexes needs to be updated regularly.

To do this, the larger search engines work automatically and analyse the Internet in order to discover new sites and update their pages.

Here again, several methods of updating are used:

- automatic updating using a web-trawling program called a "spider";
- updating of a site by the author who simultaneously labels his pages in a given engine;
- manual updating by professional surfers and cyber-journalists, whose job it is to hunt down new sites and update databases.

Each method has its advantages. Of course, the helper application can do a lot of work. Engines such as Infoseek, Excite or HotBot boast having indexed millions of pages! But power does not always mean quality, as the more pages the engine has labelled, the more specific your request must be, otherwise you will be overwhelmed by an overabundance of lists of results. Fortunately, engines can also evaluate the results provided and, in this case, can retrieve lists in order of relevance. But how to evaluate the relevance of each search engine?

Sometimes you may prefer manual selection by professionals. In addition, you may attribute great value to guides provided by cyber-journalists who give their opinion of sites visited and prepare classifications. In this way automated and human methods complement one another.

Are search engines infallible?

The answer is obviously "no". Most engines cannot interpret articles such as "a", "an" or "the". Thus a search relating to the rock group "The Who" may give surprising results!

It is therefore necessary to make the right choice, supplement your request using Boolean operators (more on these later) or search by category in order to achieve your aims.

WHAT DOES A SEARCH ENGINE CONSIST OF?

Whether you are a novice on the Web or a pro, search engines constitute your privileged point of entry to the Web. And there is no better way to understand how they work and to demystify the subject than to examine closely how they operate.

Apart from the distinction between UK-specific search engines and the more general search engines that can lead you to Web sites throughout the world, we can loosely group search engines into the following categories:

- those which spend their time indexing Web pages, such as Excite, AltaVista or HotBot;

- those which organise and grade information under headings and subheadings and constitute huge directories (such as Yahoo!, which has both worldwide and UK-specific versions);

- those which also evaluate and comment on the sites visited, such as Magellan, provided by the McKinley company.

As we have stated already, there are many techniques for constituting search engines and directories:

- automated location and indexing;

- use of manual recording by Web site designers and editors;

- recording and possibly evaluation of sites by teams of independent surfers.

Indexing Web pages and sites

There are a number of search tools which combine these various possibilities. Search engines using indexing can index millions of Web pages, e.g. more than 50 million in the case of HotBot or AltaVista, or less than 5 million in the case of engines such as Open Text or WebCrawler. In fact, current techniques allow around 3 million pages to be indexed each day. Some techniques and resources, such as those used by HotBot, even allow 10 million pages a day to be accessed. However, the number of pages indexed is not the essential indicator of an engine's quality.

Furthermore, indexing methods vary greatly from one engine to another. The criteria making it possible to differentiate the techniques used are as follows:

- regularity of updating;

- processing time of pages retrieved for indexing;

- nature of data indexed: text, images, etc.

The components of a search engine

A search engine consists of several components:

The "spider" or "crawler" is in charge of connecting with Web servers, visiting sites, reading pages and following the links on these pages within the same site. The "spider" must go to the site regularly in order to note any changes. A gigantic task!

Everything that the "spider" finds will be used to compile the index, the second component. Look at the index of this book in order see the benefits of an index. The most advanced techniques are used to design the indexes of Web sites and pages. The index is regularly updated by adding, deleting or amending entries. But there is still a time lapse between reality, what the spider has been able to detect, and this actually being recorded in the index.

The search engine, as such, is the third component! It is this which is going to analyse a request with reference to the index and prepare the results. It is also this which is in charge of sorting results by order of relevance.

Browsing and searching

Bearing in mind the difficult nature of the task (i.e. indexing sites and providing quality results to millions of surfers), no engine can claim to be exhaustive, or to perform the best, or to be infallible in the results transmitted. What is required is speed and reliability. It is easy to see that the operation of a search engine and the techniques used are a matter of compromise.

Performance varies from one system to another and depends on the nature of the search carried out. Each system has its strengths and weaknesses, and at times several engines need to be used in order to overlap and refine searches.

In general, each search engine offers:

- simple searches by keywords;
- complex searches using Boolean operators (AND / OR/ NOT) and specific commands;
- searches by subject or site category.

Depending on the performance of the helper application used, the search may be easy or difficult. You must get to know the strengths and weaknesses of each search engine, which sometimes depend on the nature of the search undertaken.

Evaluating results

A search engine's performance lies in its capacity to interpret the request, evaluate the documents found and classify search results (hits). This usually involves allocating a percentage indicating how closely the hit matches your search term, and avoids the lengthy consultation of huge lists of results. The first sites proposed are deemed to best answer the enquiry. Again, this depends on the engine correctly interpreting the request made (the search term used).

You will be able to see that search engines work fast, and even very fast most of the time, but that often results include sites or pages which are not of interest. It is up to you to sort the good from the bad!

In their defence, it must be acknowledged that search engines are not able to ask you additional questions in order to make their search more specific. It is up to you to take the initiative and if necessary word the request differently. Nor can they become enriched by your questions, although these are experiences which they could record in order to improve with time. We only know that search engines keep track of requests made in order to provide statistics which are of interest to surfers. We also know that search engines interpret the nature of requests in order to return advertising panels relating to them. In short, they are not as stupid as you might think.

You can count on the creativity of programmers to constantly include more intelligence in these helper applications.

Again in their defence, search engines are tireless workers and do not draw back from anything. You can tell them "Tourism" and they will send you thousands of pages of results. Try to ask a librarian the same thing and there is a good chance that you will be asked to be more specific, if not sent packing!

So what do search engines do to establish degrees of relevance in the results sent back to you?

First of all they use the **location/frequency method**. In plain English, they use rules which interpret the position of the words found and their frequency. The word "Tourism" will have more weight in the title of a page than in the Web page itself. It is the number of times the word appears in the page itself which will eventually decide between two pages where the keyword appears in the title.

The position of a word in the page is also relevant. Thus, the engine can determine whether the word in question appears at the top of the page, in a title or in the first paragraph, all of which will have more weight than its appearance at the bottom of the page.

Most engines use this basic method with their own variant. That is why, amongst other things, different engines will produce different results for the same request.

As we have said, Internet search engines are far from equal:

- some record more pages than others;
- some update their index faster; and
- their engines are of varying efficiency.

Other methods of differentiation are used by search engines. Thus, WebCrawler tests the **popularity of links**. It will give preference to a page to which lots of links point rather than a page which has fewer links.

One might also imagine that search engines associated with qualified site directories will give better marks to manually-evaluated sites. A site visited by a professional team will be preferred and allocated a better mark than a site unknown to the team.

Thus far we have made no mention of the interpretation of meta tags. What are these? Meta tags are hidden programming codes which are used by programmers to describe a Web site or page and record the associated keywords. These commands in HTML language are used because engines interpret them. So far so good, since meta tags were dreamed up precisely to facilitate indexing and further searches. It is by using these commands artfully that the programmer will be able to ensure that his page is "retrieved" by search engines.

While some search engines, such as Excite, ignore the keywords recorded in meta tags, others, such as HotBot, make great use of them. But the drawback is that a programmer can lead the search engine into error. As a result of trying to ensure that his page is visible, his programming may prove to be a deception. An amusing example to show what I mean. Starting from the assumption that there is great demand for erotic sites, a programmer will not hesitate to code the word "nudity" in a meta tag in order for his page to be visible to those in search of such sites – even if the site contains nothing more than pictures of classical Greek statues!

There are engines of course which try to cut out this type of abuse. Thus, if a word is repeated a hundred or so times one after the other in order to win the "frequency" competition, the engine may decide to pass over the site. The programmer's trick will not be rewarded, and we can but be glad about this.

Since meta tags are absolutely freely defined, it is inevitable that engines will make mistakes in their interpretation. While the Excite solution is protective, it may penalise strict authors. The ideal solution would be to create a description standard accepted at the highest level, a standard on which engines could depend. Doubtless this will come. In the meantime, cyberspace, like the real world, is an imperfect world.

The designers of Web pages and sites have several ways of getting themselves known by indexing applications:

* *site title;*

* *site description: this is a hidden programming code (a tag in HTML language) containing the site description given by the editor or independent author; or*

* *keywords: this is another hidden code which contains a list of keywords attributed to sites.*

Today, everyone is free to describe a site as he chooses, and in the long run this is what can create anomalies in "hits" (search results).

For the curious, this is how meta tags are coded in HTML language:

```
<META NAME="keywords" CONTENT="cuisine,
gastronomy, soil, wine">

<META NAME="description" CONTENT="A
magazine on traditional cooking, good
wine and the fruit of the soil.">
```

And for engines that ignore meta tags there are still comments:

```
<!--// A magazine on traditional cooking,
good wines and the fruit of the soil. /
/--!>
```

Presenting results

Indexing pages and analysing requests are the basic tasks of the search engine but it still has to produce the results as comprehensively as possible for the surfer. Here too, methods vary from one engine to another. Every kind of site description corresponding to a request may be displayed:

- the descriptions incorporated in meta tags;
- Web page titles;
- the first lines of a Web page;
- the manual summary made by a specialist surfer or cyber-journalist;
- the page address (URL);
- the page size;
- the date of the last update;
- the degree of relevance.

The site address, or URL (Uniform Relation Locator), is also used in searches. While it is not the major point of entry of engines, a search by URL may be proposed by some of them. Thus addresses including keywords may be searched for or a search may be restricted to a specific domain.

Remember that a Web site address is structured as follows:
http://www.server_name.domain/site_name/page_name

- ***http**. This is the communication protocol used, in this case HyperText Transfer Protocol; this protocol was designed to interpret hypertext mechanisms (links between pages) developed on the Web.*

- ***www.*** *Designates the World Wide Web, the most populate area of the Internet. Developed in 1990 by CERN (Centre d'études et de recherche nucléaires) in Geneva, the Web makes use of hypertext technology, i.e. the connection of pages and documents using links. On the Web, one page contains all kinds of link:*

  - ***Links in the same page****. These are also called anchors. They make it possible to move quickly around a large text document (one page).*

  - ***Links with another page of the same site****. These links are used to move from one page to another within the same site or service. They allow the user to navigate the site.*

  - ***Links with a page of another site.*** *These links are offered by the developers of Web sites to reference other useful sites which may, of course, be on other servers (computers storing Web pages). It is possible in this way to move from one site to another, from one Web server to another and often from one country to another by just a few clicks of the mouse. The user navigates without worrying about where data is located, all that matters to him is what he is looking for.*

It is this natural method of navigating using hyperlinks which is commonly known as "surfing the Internet".

SEARCHING THE INTERNET: PRINCIPLES

While some engines offer you options (choice from a drop-down list, boxes to be checked), others leave you free to formulate your requests. In the first case your searches are guided, in the second you are left to yourself. Remember that all engines have help files explaining how they work. From these files you will learn in particular the authorised commands and syntax, which are useful in making sophisticated searches.

As a general rule, the commands available are not displayed. Although there are not many of them, their use is not always standardised, which should encourage you to read the following explanations.

Match any or Match all?

When you type several keywords in the search box, two things may occur:

- either the engine searches for the pages where all the cited keywords appear (HotBot, Lycos, Infoseek); or
- the engine indicates the pages where at least one of the words appears (AltaVista, Excite, WebCrawler).

In the latter case, the number of words found may be one of the relevance criteria. The engine will display at the top the pages where it has found all the words. For the others it will indicate the number of words found, e.g. 2/4.

Bearing in mind these differences in behaviour, this is how each of the families of engine manages to respond to requests:

- to obtain "Match any" with HotBot, for example, you must use the menu command options or code the request using the OR operator: "Tourism or England";
- to obtain "Match all" with AltaVista, for example, you must use the operators + or AND: "Tourism + England" or "Tourism AND England".

Boolean operators

Boolean operators are used to express logical search conditions. The two basic operators are AND and OR. While the AND operator is unambiguous, OR can take two forms:

- the normal OR: in A OR B, the result contains A, B and "A and B".

- the exclusive OR: in A OR B (exclusive OR), the result contains either A or B. The result "A and B" is excluded.

Only the normal OR is used by default in search engines.

Exclusion

All engines allow you to exclude a word from results. If you are interested in tourism outside the UK, you can exclude the word "UK". Nothing complicated, except that the syntax varies from one engine to another.

- the – sign is used with AltaVista or Lycos;

- AND NOT is used with Infoseek, HotBot or AltaVista;

- NOT is used with WebCrawler or Lycos;

- BUT NOT is used with Open Text;

- menu commands may also be used with Excite or HotBot.

This simple example should encourage you to be vigilant in formulating requests.

It is very simple to formulate a request using a search engine, and the reply does not take long to arrive. But it is very simple to go astray and interpret the reply with respect to the request which you have in your head, whereas what you asked was something else completely!! A good tip is to re-read your request carefully and check how the search engine works.

Searching for sentences

The command "Tourism and UK" does not have the same effect as the command " "Tourism in UK" ". In the first case, a search is carried out relating to 3 keywords, probably with non-interpretation of the word "and" by the search engine. In the second case, pages including precisely the sentence "Tourism in UK" are searched for.

Most engines use inverted commas for this, or offer a choice via a menu (Excite, HotBot or Open Text).

Proximity

Some engines allow you to indicate parameters relating to proximity between the keywords designated. For example, A NEAR B indicates that word A must be close to word B (used by AltaVista, Lycos, WebCrawler or Open Text).

Note that Infoseek offers the command [A B]: words between square brackets.

You can even specify the level of proximity:

A NEAR/10 B indicates that A and B are within 10 words of each other (used by Lycos and WebCrawler).

Wildcards

Wildcards, familiar to DOS users, are probably new to others.

This is how they work in general:

- the * character indicates a string of characters of any length. "Art*" thus makes it possible to search for art, artist or artisan.

- The $ character indicates a character in a specific position. "Artist$" carries out a search for artists (in the plural) or artiste.

Searching by field

Some engines offer particular searches by type of field:

- the title of Web pages;
- addresses or URL;
- domain names;
- links.

This is the case with AltaVista, HotBot and Infoseek in particular.

> *Be careful regarding:*
>
> - *capitals;*
>
> - *the use of Boolean operators and commands specific to each engine;*
>
> - *the position of brackets in writing complex logical operations;*
>
> - *typing errors!*

GETTING STARTED WITH YOUR NAVIGATOR

Recording a search engine on a home page

Your Internet browser (Microsoft Internet Explorer or Netscape Communicator, the two most frequently used tools) makes it possible to parameterise the initiation of a Web site when you open a session. As most of your Internet connections start *a priori* with searches for information and sites, you might as well reference a search engine as start page.

This is how to proceed with Internet Explorer 5.0:

1. Click on Tools.

2. Click on Internet Options.

3. Select the General tab, if necessary.

4. In the Home page box, enter the Internet address of the selected engine.

5. Confirm by clicking on the Apply button, then on OK.

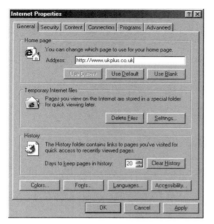

Figure 1.1: Recording the UKplus search engine as start page in Internet Explorer 5.0

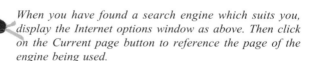

When you have found a search engine which suits you, display the Internet options window as above. Then click on the Current page button to reference the page of the engine being used.

Search tools integrated in browsers

Netscape (Communicator) or Internet Explorer have a Search button which gives access to specific search resources on the Internet.

This button in fact gives each of them access to a specific Web page which offers links with pre-selected engines.

Searching with Netscape

The Infoseek search engine is pushed to the fore on the Netscape Netcenter search page (**http://home.netscape.com/escapes/search/netsearch_2.html**).

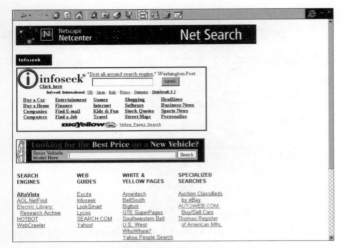

Figure 1.2: The Netscape Netcenter home page

Four other categories of search tool are also offered:

- search engines in the strict sense, like AltaVista, AOL NetFind, HotBot or WebCrawler (see Chapter 3);

- Web guides (searching by category): Lycos, Yahoo! (see also Chapter 3);

- yellow or white pages in order to search for companies or people (address, telephone and e-mail): Bigfoot, WhoWhere (see Chapter 5);

- specialist services: sale by auction, purchase and sale of vehicles.

You will also find access to a subject guide on the home page: Netscape Guide by Yahoo! **(http://guide.netscape.com/guide)** produced in collaboration with Yahoo!

Figure 1.3: The on-line Netscape guide

▰▰▰ Searching with Internet Explorer 4.0

By clicking on the Search button of Internet Explorer 4.0, you open an exploration phase in the right hand part of the screen.

Just type the keywords and check the search engine of your choice from among the 4 offered, namely:

- Yahoo!;
- Lycos;
- UK Plus;
- Yell.

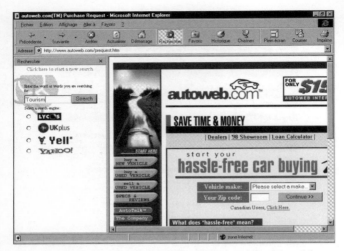

Figure 1.4: The exploration phase of Internet Explorer 4.0.

Figure 1.5: Formulation of the result with UK Plus

While the first two engines are widely renowned, the second two, UK Plus and Yell (two UK-specific search sites) are the result of more recent agreements with Microsoft.

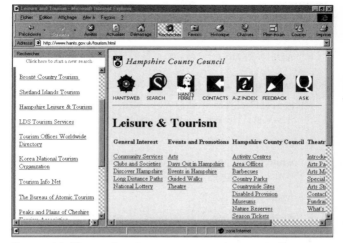

Figure 1.6: The exploration phase, with the list of hits, with one of the sites displayed on the right

The list of sites is displayed in the exploration phase. By clicking on one of them you can display the content of the site selected in the right hand part of the screen.

To initiate a new search (other keywords, change of search engine), just click on the words "click here to start a new search" in the exploration phase.

At any time just click on the Search button to close the exploration phase and devote all available space to the site visited.

Microsoft also offers a search page: Search the Web **(http:// home.microsoft.com/access/allinone/asp)**. Five general search engines are offered from this page, namely:

- Lycos;
- Yahoo!

- Excite;
- Infoseek;
- AOL NetFind.

To this must be added the following seven sections:

- general search of the Web, with other tools such as AltaVista, HotBot or WebCrawler;
- guides and directories such as NetGuide, Magellan or Top 5%;
- white pages for finding people and addresses: WhoWhere, Four11, Bigfoot;
- finding forums;
- finding real-time chat sessions;
- international searching using Infoseek International; and
- specialist search tools: cards, hotels, music, cinema, finance, etc.

It is in this last category that we find a further two possibilities:

- Microsoft: to find information published on the Microsoft site.
- Encarta: for searching in the Microsoft general encyclopaedia.

In each of these sections it is possible to activate the tool of your choice. By checking the box with the mouse you display the search box of the selected search tool at the top of the page.

You thus have access to 34 general or specialist search tools. This selection is not fixed and can change depending on the partnerships established by Microsoft.

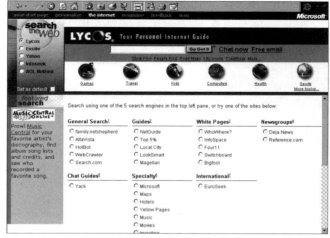

Figure 1.7: The Microsoft search page, with the default display of the Lycos search zone

Figure 1.8: Searching the on-line Encarta encyclopaedia

To change your choice of default engine, just reference a new tool and check the box "Set as default".

10 tips for successful searches:

- go from general to specific: use a fast search before refining the request;

- select the right tools, according to your requirements;

- use the well-known search engines in the first place;

- *use UK-specific search engines like UK Plus or Yahoo! UK and Ireland to find UK Web sites;*

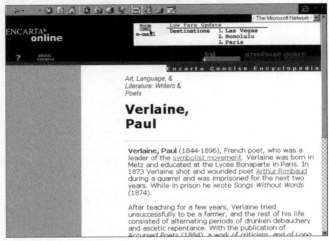

Figure 1.9: A hit on Encarta online with the keyword "Verlaine"

- *overlap and possibly enrich results using more than one search engine;*

- *examine carefully the search options offered by the search engine;*

- *use search-by-keywords and search-by-site categories together;*

- *if necessary, use advanced search techniques (power searches) and the search engine's help file in order to understand how it works;*

- *use metasearch engines for specific searches, reducing your work time.*

Chapter 2

UK Search Engines

THE CONTENTS FOR THIS CHAPTER

- Find the UK versions of popular worldwide search engines
- The best UK-specific search engines

For everyday Web surfing and general research, there are large numbers of search engines available that can help you find what you're looking for. But there are times when only a UK site will do. For example, if you need legal or financial advice, want to book theatre or rail tickets, or want to check tonight's TV schedules, international search engines may not be much help. Instead, you want a search engine that's guaranteed to answer your query by providing links to UK web sites.

There is a good choice of UK search engines, falling into two categories:

- UK versions of international search engines (Yahoo! UK and Ireland, for example);

- UK-specific search engines, listing only UK sites and ignoring the rest of the Web.

UK VERSIONS OF INTERNATIONAL SEARCH ENGINES

"UK version" here means the adaptation of an international search engine (see Chapter 3) to favour UK sites. Popular search engines and directories such as Yahoo!, Lycos, InfoSeek, and Excite each offer a UK version. In most cases you can choose whether to search the entire Web or just the UK portion of it, so you may choose to use the UK-specific version in preference to the international version, whatever type of information you want to find.

▬▬▬ Yahoo! UK and Ireland

http://www.yahoo.co.uk
More than a search engine, Yahoo! UK and Ireland is a subject directory which offers searching by keywords. It does not provide its own summaries of the sites found, but instead uses (usually very brief) site descriptions submitted by the author of a site.

On the main Yahoo! UK and Ireland page, the search choices are simple:

- All sites: searches the entire Yahoo! directory. UK site matches will appear at the top of the list of search results with international sites below;

- UK & Ireland sites only: limits the search to sites in the UK and Ireland.

Along with the keywords you want to search for, you can use inverted commas to indicate a sentence or expression, and + or - signs to include or exclude a word.

To gain a little more control over the way the search is conducted, click the Options link beside the Search button. From the Options page you can make more detailed selections:

- search Yahoo! (for web sites), Usenet newsgroups or email addresses;

- search for direct links to web sites that match your keywords or for matching Yahoo! categories;

- choose to search for the exact phrase you entered, matches on all words, matches on any word, or a person's name.

On this page, you can also choose the number of hits displayed on each page of the search results, and limit the search to include only recently-added sites.

Yahoo's search results are presented in two sections:

- Site categories: this is the subject hierarchy used by Yahoo!. Each category contains a number of web sites, plus further categories that form a subset of that category (for example, Society & Culture contains a Relationships category, which itself contains a PenPals category).

- Web sites: these are direct links to web sites that match the keywords you entered. UK and Ireland sites are marked with flags to distinguish them from international sites.

Figure 2.1: The Yahoo! UK & Ireland logo

Figure 2.2: The main Yahoo! UK & Ireland search page

Figure 2.3: Search results from Yahoo! UK & Ireland

Lycos UK

http://www.lycos.co.uk

Like Yahoo! UK and Ireland, the front page of Lycos UK offers a straightforward choice between searching the entire Web or limiting the search to UK and Ireland sites. The search engine returns a list of hits with a degree of relevance expressed as a percentage.

Lycos offers a range of search services on its Search Options page:

- search the entire web;
- search for UK and Ireland sites only;
- search for pictures, sounds, books or MP3 audio files.

This page allows you to customise your search in just about any manner you could think of. You can also run searches for UK businesses, jobs and classified advertisements or take advantage of Lycos' translation service.

For each search made, Lycos keeps track of each word on each Web page listed. Common words such as "the", "a" and "and" have been deleted, allowing the search to be optimised by eliminating words which are not significant. All search engines work in a similar way. However, searching for an expression such as "To the one" may cause problems.

Figure 2.4: The Lycos logo

Figure 2.5: Customising your search at Lycos UK

Infoseek United Kingdom

http://www.infoseek.co.uk

Infoseek is another large international search engine. Along with the usual options to search the entire web or just the UK portion of it, a drop-down list allows searching in more than a dozen other countries. Other useful options include a Film Finder and Yellow

Pages, courtesy of Yell (the traditional Yellow Pages we all know and love, which has its own site at **http://www.yell.co.uk**).

In addition to the traditional search options, Infoseek lets you compose sophisticated requests using the following commands:

- link: pages including an Infoseek link;
- site: search for sites;
- URL (address): search for web site addresses;
- title: search for document titles.

As an example, entering "URL:science" will start a search for documents whose address contains the word "science".

Hits include a degree of relevance expressed as a percentage and the document date.

Figure 2.6: The Infoseek United Kingdom logo

Figure 2.7: Searching with Infoseek United Kingdom

Excite UK

http://www.excite.co.uk

Excite is an original, powerful search engine which encourages searching by ideas and concepts rather than keywords. It uses the ICE technique (Intelligent Concept Extraction) to determine the correspondence between words and ideas. Hits are sorted according to their relevance, with the pages that best match the search criteria placed at the top of the list.

When the search results appear, a site that corresponds exactly to what you were hoping to find can be used as the basis for a more specific search. To do this, just click the "More Like This" link beside the URL. The search engine will then use this document as a reference for another search on similar sites.

Excite UK lets you search the entire Web, UK sites only, European sites or Usenet newsgroups. Check the appropriate radio button below the search field. For more precise searching, you can use the + sign to indicate words that must be included in hits and the - sign to exclude a word from the search. You can also use inverted commas to indicate phrases or sentences.

Using Boolean functions with Excite

Boolean functions override Excite's concept-based searching and allow you to run searches for the exact keywords you specify. The Boolean operators available are AND, AND NOT, OR and brackets. These operators must be entered in upper case, with a space before and after them.

AND: The documents found must contain all the words linked by the AND operator.

OR: The documents found must contain one of the words linked by OR.

AND NOT: The documents found must not contain the word which follows the term AND NOT.

( ): Brackets are used to join together segments of Boolean inquiries to allow even more specific searches. For example, to find documents containing the word "fruit" and either the word "banana" or the word "apple", you would type "fruit AND (banana OR apple)".

The hits produced by the Excite search engine include the percentage of relevance, the site or page title, the URL and a summary.

Along with its UK version and the international site at www.excite.com, Excite offers version adapted to other countries and languages including Australia, China, Germany, Japan, Italy, France, Sweden and the Netherlands.

Figure 2.8: The Excite UK logo

Figure 2.9: Choose whether to search the entire Web, UK sites, European sites or newsgroups

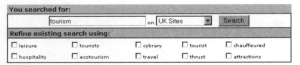

Figure 2.10: Selecting words to add to an Excite search

LookSmart United Kingdom

http://www.looksmart.co.uk

LookSmart is a directory-based search engine that uses "categories" in much the same way as Yahoo. LookSmart comes from the same stable as the popular and powerful international search engine AltaVista, and it's actually this engine which carries out your searches.

The LookSmart search interface is fast and simple: just type in your keywords and click the Go! button. Like Yahoo, LookSmart presents links to matching categories first, followed by links to web pages that match your search terms. The web page results give the title of the page or site, a brief description and a link to the LookSmart category in which that site was found, allowing you to browse the category for links to similar sites.

Figure 2.11: The LookSmart logo

Figure 2.12: Search results from LookSmart

If you prefer browsing by category to running keyword searches, LookSmart offers an "Explore" mode that makes this easy. Click a major category from the list at the left of the page and a list of subcategories will appear beside it. Click a subcategory to see a third level of categories, and so on. After a few clicks, a page of links to sites related to a specific category will open.

Figure 2.13: Moving through subject categories using LookSmart's "Explore" mode

UK-SPECIFIC SEARCH ENGINES

This section takes a look around search engines specifically built to provide links to sites in the UK. If you know that the information you want can only be found on a UK site, these search engines make an excellent starting point.

▄▄▄▄ UK Plus

http://www.ukplus.co.uk

UK Plus is a category-based directory site, in a similar vein to Yahoo!. Its great strength is that every site listed has been reviewed by the folk at UK Plus. As this is obviously a time-consuming business, UK Plus is a good deal smaller than traditional robot-built search databases. However, the immediate benefit is that you have a good idea of what a site will contain before you click the link to visit it. Robot-built databases pick up the bad sites with the good, and what you find when arrive at one of those sites may differ enormously from what the site's description led you to expect!

As with Yahoo! and LookSmart, you can search the directory by selecting a major category from the UK Plus home page and drilling down to more specific sub-categories. For a faster automated search, use the text field and buttons at the top of the home page. You can also run an Infoseek-powered search of the entire Web by clicking the radio button marked "all of the web".

Figure 2.14: The UK Plus logo

Figure 2.15: The simple search form and category list at
UK Plus

Yell

http://www.yell.co.uk

Yell is the UK's Yellow Pages in its online incarnation. Yell offers
a range of useful services:

- Yellow Pages: search for UK businesses;

- Web: a searchable directory of UK web sites;

- Entertainment: find out what's showing at your local cinema
  or book theatre tickets;

- Shopping: find secure UK online shopping sites in a number
  of categories;

- Travel: find flights, hire cars, holidays and accommodation.

Searching for businesses couldn't be easier. Just enter a company
name (if you're looking for a particular company) or a business
type, and type in the location of the business. The search results

list company names, addresses, telephone and fax numbers. Where more information about a company is available, or the company has a web site you can visit, you can click a link to find out more.

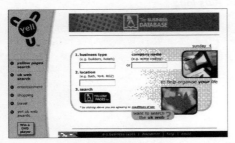

Figure 2.16: Searching for businesses at Yell

Yell's web searching is very similar to that of UK Plus: you can delve into the categories yourself or run a simple keyword search. Like UK Plus, Yell researches and describes all sites by hand to give you a clearer idea of what to expect. Yell will only display the first 25 matches found, so it's important to pick your keywords carefully to ensure that you really do specify what you're looking for.

To find out which sites most Yell users have been jumping to in the past week, click the Top 30 link at the bottom of the window.

Figure 2.17: The search services and web site categories at Yell

▰▰▰▰ Rampant Scotland Directory

http://users.rampantscotland.com

What it is that makes Scotland rampant isn't explained, but this comprehensive searchable directory makes an ideal starting point to finding Scottish sites and information. The large number of categories (and relatively small number of sites) means that a few clicks on the category titles should quickly lead you to what you want. If you prefer, you can run a keyword search of Rampant Scotland, or a metasearch of the entire Web (see Chapter 4) to combine hits from five major international search engines.

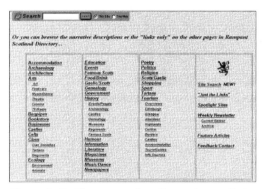

Figure 2.18: The comprehensive category list and simple search form at Rampant Scotland

▰▰▰▰ UK Index

http://www.ukindex.co.uk

For surfers in a hurry, UK Index provides a "quick search" form on its front page. But a click on the "search page" link or a direct jump to **www.ukindex.co.uk/uksearch.html** reveals a lot more! UK Index cleverly combines the directory structure (used by Yahoo!, UK Plus and others) with a keyword search form. This allows you to combine a keyword search with the Boolean operators AND, OR or PHRASE, and restrict your search to particular categories.

Start by entering either one or two keywords and selecting the appropriate operator (this is ignored if you enter only one keyword). Then check the boxes beside the categories in which you want to search. UK Index returns only those sites which are found in all the categories you chose. For example, if you search for "mice", selecting the Nature category will ensure that the results don't contains links to computer accessory sites.

Figure 2.19: Combined keyword and category searching at UK Index

▋▋▋▋ Mirago

http://www.mirago.co.uk
http://www.mirago.co.uk/zone

The Mirago search engine has several features that recommend it, but one in particular stands out: clicking the button marked "Kids Only" will take you to the second of the URLs above – a 'family-friendly' version of the search engine. This version of the Mirago engine uses a combination of automatic and manual filtering to remove links and site descriptions that you wouldn't want your kids to see. While these filters are not entirely infallible, they do give a good reason for adding the 'safe' version of this site to your browser's Favorites or Bookmarks lists.

The search options themselves are equally comprehensive on either version of the site, allowing you to search for UK sites, or sites containing images, sounds, video or multimedia. Simple Boolean operators can be selected from the drop-down "Look For" box:

- all of the words;

- any of the words;

- a person's name;

- words as a phrase;

- words near each other in the document;

- resources using Boolean.

Selecting the last of those options allows you to use advanced syntax to refine your search. The operators AND, OR, NOT and NEAR can be used, and phrases can be specified by enclosed them between double-quotes.

You can also enclose expressions in brackets, as with Excite, and add an asterisk wildcard to the end of a word (for example, a search for "south*" would return hits on "southern" and "Southampton").

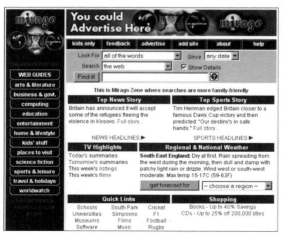

Figure 2.20: Mirago's "family-friendly" version offers safe but comprehensive searching

Chapter 3

Worldwide search engines

THE CONTENTS FOR THIS CHAPTER

- Discovering the heavyweight search engines
- Techniques
- Functions and services offered by the different search engines
- Which language dominates the Web?
- Which search engine to choose?

WORLDWIDE SEARCH ENGINES

These are the giants of the world of search engines, being both the oldest and the largest. Most of them catalogue millions of Web pages. Enough to satisfy the most curious. But we still need to learn how to master them, as they do not balk at sending you thousands of replies to most queries. Therefore it is up to you to make your request specific using advanced search functions allowing you to combine criteria and to filter (exclude) results. You will now discover them in their worldwide English version.

▬▬▬ AltaVista

http://www.altavista.digital.com/

The search engine of the computer manufacturer Digital Equipment is one of the leaders on the Internet. A number of search tools, including Yahoo! and Bigfoot, also use AltaVista technology. This was developed in the research laboratories of Digital Equipment at Palo Alto. It was in 1995 that AltaVista first appeared on the Web, with the ambition of cataloguing the entire Internet! A catalogue which represents around 200 Gb in its global version, and which will be polled in less than one second in most cases!

AltaVista thus offers a global search of the Web or Usenet (forums). It is possible to search in the language of your choice. By default, the search engine is set to Any language. It is possible to indicate your preferences to the search engine, enabling you to save time.

The advanced search mechanism will be used when, for example, it is necessary to use Boolean expressions or a document validity range (dates).

You can carry out textual searches using inverted commas, or specific searches using the search engine's commands:

Domain. Search of a particular domain (.com, .org, .fr, etc.).

Image. Search for photographs, for example.

Link. Search for links.

Also on the programme for AltaVista, searching for people (by name) and companies (business, location, etc.).

 In normal search mode, AltaVista sorts hits by order of relevance. This mechanism is no longer active in advanced search mode. It is up to you to activate it if necessary, using the Ranking text box, otherwise hits will be produced in any order.

Searching by date
It is possible to filter hits (Web pages) based on the date of documents. To do this you must complete the FROM and TO text boxes. Dates must be completed in the dd/mmm/yy format, where dd and yy represent days and years and mmm represents the first three letters of the month.

Figure 3.1: The main search box of AltaVista

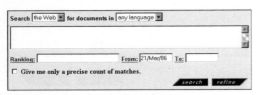

Figure 3.2: The choice of preferences (language, search type, presentation of results)

Figure 3.3: Advanced searching with AltaVista

Figure 3.4: The scrollbar in hit pages. Here, 17 pages of 10 hits are to be displayed.

Since July 1997, AltaVista has offered searching in 25 different languages. It uses a multilingual dictionary to determine the dominant language of each page indexed. It is thus possible to display the hits corresponding to the language of your choice. This is much more efficient that searching by domain.

▄▄▄▄ Excite

http://www.excite.com/

Another giant, since Excite updates an index of 50 million Web addresses (URL). This search engine is very powerful, offers searching by category and keyword, and produces hits in order of relevance.

Excite is original in more than one way and has a very useful function called *Excite Search Wizard*. At the end of a search, the Wizard suggests a list of words that you can then add to another search. To do this, just check one or more words in the list of suggestions.

The words proposed are derived from words entered in the search text box and sent to the search engine. Excite then makes a list of similar words or concepts to help you to express your requests.

Thus, a search relating to the word "kennedy" will produce a list of suggestions including "assassination", "assassinated", "jfk" and "onassis", words which make reference to the life (and death) of the famous US president. The list of suggestions will also include words such as "shuttle" (space craft), "launch", "flight" and "nasa", which this time relate directly to the *Kennedy Space Center*.

Take note of Excite's suggestions; they can be of real assistance in formulating your requests.

This highly practical mechanism is not yet available in Power Search, but will be in the future. In the meantime, it is quite sufficient to use it in normal search mode.

Excite also offers specialist services such as:

- searching for maps with City.net;

- searching for people (People Finder); and

- searching for companies in the United States (Yellow pages).

Figure 3.5: The Excite home page

Figure 3.6: The result of searching with Excite

Yahoo!

http://www.yahoo.com/

Here is another heavyweight search engine. More of a guide-directory than a traditional search engine, Yahoo! operates according to the same general principles for the user. It sends categories and sites corresponding to the request. Thus you will go through the categories suggested and catalogue the individual sites found.

Yahoo! also offers a current events search service: News Articles and Net Events.

Yahoo! searches in four directions (or four databases):

- **Yahoo! Categories.** The categories and subcategories constituting the highly detailed classification of the search guide.

- **Yahoo! Web Sites.** Web sites in the strict sense.

- **Yahoo!'s Net Events & Chat.** Live events and chat.

- **Most recent News Articles.** Current events items.

Yahoo! sends the categories referenced, then the sites indexed corresponding to your request. If no correspondence is found, Yahoo! must then undertake a full text search on the Web using the AltaVista search engine as backup.

Yahoo! evaluates the results of a request according to three main rules:

- documents containing the most keywords are the most relevant;

- documents whose title contains keywords take priority over those where the words searched for appear only in the body of the Web page or in its address; and

- the most general categories (at the top of Yahoo!'s Internet hierarchy) are better ranked than more specific categories (and therefore at the bottom of the hierarchy).

Yahoo! also offers all sorts of additional services, such as:

- searching for people;
- searching for maps;
- searching for companies.

And also more original services such as:

- **Visa Shopping Guide (online buying).** A search tool for online shopping.
- **Yahooligans for Kids.** A search engine dedicated to sites for children.

Figure 3.7: Searching with Yahoo!

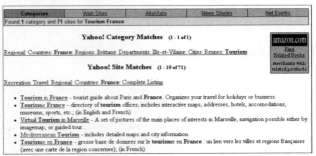

Figure 3.8: Presentation of results - categories and sites

Figure 3.9: A search tool for sites for children

HotBot

http://www.hotbot.com

Created by the famous American magazine *Wired*, HotBot also plays in the court of the large search engines.

Figure 3.10: Searching with HotBot

A request may relate to:

- all keywords;

- some of them;

- the exact phrase;

- the words in the title; or
- conditions expressed by Boolean operators.

All kinds of criteria may be specified when searching, such as:

- the recent nature of information; and
- the location of data and domains.

To this must be added a highly practical function making it possible to indicate the type of media searched for:

- images;
- sounds;
- video; or
- Shockwave animations.

The HotBot search engine also allows other searches:

- e-mail;
- forums;
- shareware;
- current events; or
- small advertisements by category (car, property, computer, jobs, etc.).

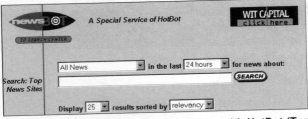

Figure 3.11: Searching for current events with HotBot (Top News sites)

Please choose a category:

Vehicles
Cars, Trucks, Vans,
Motorcycles, RVs,
Parts, more...

**Rentals &
Roommates**
Apartment Rentals,
Roommates, Vacation
Rentals, Timeshares

Personals
Find your ideal match
or a new friend.

**General
Merchandise**
Stereo/Video, Sports,
Furnishings,
Photography, more...

Everything Else
Pets, Aircraft,
Announcements,
Watercraft,
Tickets/Events,
more...

**Computers/
Software**
PCs, Macs, Drives,
Monitors, Printers,
Memory, more...

Employment
Over 75,000 Career
Jobs.

Real Estate
Nationwide listings

**Opportunities &
Services** NEW!
Business
Opportunities and
Services Offered

Figure 3.12: Searching small advertisements using HotBot

For advanced searching, click on the Supersearch link. A detailed
form appears. You can thus enter:

- additional words or phrases;

- the date or validity range for pages;

- where to search (continent); and

- the media to be found on pages (images, sounds, animations,
  etc.). You can even state the type of file searched for by
  indicating the extension!

Searching the Internet

Figure 3.13: Advanced search using HotBot Supersearch

Magellan

http://www.mkkinley.com/

The strength of Magellan compared with the other search engines is the comment made on most sites. The analysed sites are also awarded a mark. You can restrict the search to sites which have been visited and reported on by the McKinley company's team of surfers. You can even stick to those which have been awarded the highest marks ("Green Light" sites). Although you obtain less addresses when making a request than with Yahoo! or AltaVista, you already have an idea of the content and quality of the site before going to visit it. The time saving is considerable.

Magellan uses the technique of *Intelligent Concept Extraction* (ICE) in order to identify the relationships between words and ideas or concepts. The results are then sorted according to relevance.

Figure 3.14: The Magellan home page

The "Find Similar link" function will be used based on one of the URLs returned. This will make it possible to restart the search from an initial hit which is deemed worthwhile, and thus to enrich the previous search.

Like all large tools, Magellan also offers additional services:

- searching for people and e-mail addresses;

- yellow pages (companies); and

- miscellaneous services (sports results, stock exchange prices, weather reports, etc.).

WebCrawler

http://webcrawler.com/

WebCrawler is a subsidiary trademark of Excite, so may not have a long-term future. However, it exists, and it offers traditional searching by keywords, which gives good results.

WebCrawler also works in natural language (English). Therefore it allows a request to be expressed simply, without worrying about Boolean operators!

If the search function is not sufficient, you can use the guide part, with an allocation of more than 5,000 sites per category. The editorial team regularly adds and deletes sites from this catalogue, which thus records the best of the Web.

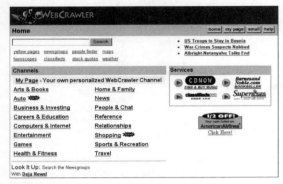

Figure 3.15: The WebCrawler home page

In addition, WebCrawler offers the following search services:

- yellow pages;
- forums;
- searching for people (see Chapter 4);
- maps; and
- small advertisements.

Open Text

http://www.opentext.com/

Open Text presents lots of links on its home page, and in particular folders and current events relating to data processing techniques. An interesting front door for enthusiasts. The search text zones are at the foot of the page! They allow you to ferret through the site, but also the Web. It is possible to use the URL **http://index.opentext.net/main/simplesearch.html** to go directly to the Web search functions.

Originally a document manager, Open Text was designed in 1988, well before the Web appeared (end of 1990). The technique was therefore used on the Web to index pages. Open Text thus references more than 5 million pages (only one tenth of the number indexed by the leaders), but all the same this represents 170,000 sites! The Open Text technique is also marketed for companies under the Livelink Intranet label. It constitutes an efficient solution for managing pages on an intranet (local network or interconnection of private networks based on the Internet technique). So it is a useful search tool.

A drop-down menu allows you to choose between three search options:

- the exact phrase;
- all words stated; and
- at least one of the words mentioned.

Worthy of note are Power Search for sophisticated searches, Current Events for consulting current events in the American media, or Newsgroups for ferreting through forums.

Finally, we would point out that it is possible to use Open Text in Japanese, subject to parameterising your browser so that it correctly displays the characters of the Japanese language.

Figure 3.16: The very full home page of Open Text. Use the lift to display the search engine!

Infoseek

http:/www.infoseek.com
Another leader in terms of surfing the Web, Infoseek offers all the major features of a large search engine. So we will spend a little time looking at its most original services.

One of these is the personalised news service. Choose the subjects which interest you from among the 13 suggested, and for some of them enter one or more keywords. The current events found are sorted by date initially, and then by degree of relevance.

Infoseek can also be installed as a Web channel for Internet Explorer 4.0 (see Chapter 12). This time you no longer look for information, information comes to you.

Once more, there is access to online dictionaries (under Reference) such as the Webster Dictionary (definitions and spelling checks) or Roget's Thesaurus (for English synonyms).

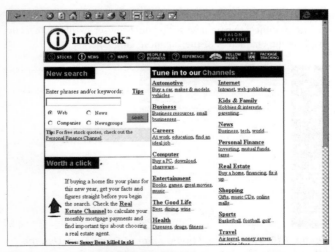

Figure 3.17: The Infoseek home page

WHICH LANGUAGE DOMINATES THE WEB?

AltaVista enables you to check the weight of each language on the Web. To do this you must use the Advanced Search mode. Enter the character * in the search text box and click on Search.

You then obtain the number of pages of the index, i.e. N, all languages mixed up.

Do the same thing again, selecting a language. You obtain a number N1. N1/N gives you the percentage presence of this language on the Web.

At the time of writing this book, the situation was as follows:

Language	Number of pages	Proportion
English	74,975,881	85.24%
Swedish	1,673,810	1.92%
French	1,211,489	1.41%
German	988,229	1.14%
Italian	663,507	0.76%
Spanish	571,507	0.66%
... out of a total of almost 87million indexed pages!!!		

Worthy of note is the excellent position of Swedish. Japanese, Chinese, Finnish and Portuguese come after Spanish.

WHICH SEARCH ENGINE TO CHOOSE?

Yahoo! is a directory more than a traditional search engine, but it has already earned its stripes. The wealth of its organisation and rigour of classification make it a valuable tool. This hierarchisation of categories aids understanding. It is a good starting point for anybody who is unfamiliar with the Web. If Yahoo! does not find an answer to your request, it hands over to AltaVista.

AltaVista is the Web's heavy artillery. Vast and powerful, this search engine is useful for scanning the Web thoroughly. You need to give yourself time to go through the results. AltaVista has a natural language search function which is worth looking at.

Excite is one of the market's dead certs. It is a search engine with enormous intelligence. It tries to interpret your request and suggests, as we have seen, similar words and concepts. It has many fans amongst surfers and will also be a very useful tool for you.

HotBot is an original search engine. It performs excellently and is also very much appreciated by surfers from day 1. Its relationship with the magazine Wired has increased its popularity amongst surfers. It is also well used by professionals.

Magellan is also a directory. As a bonus, it offers comments and evaluations given by highly competent editorial teams. With Magellan, man takes precedence over software, which is no bad thing.

WebCrawler is not as popular for purely strategic reasons. The operator America On Line is replacing it with Excite, and it bears the mark of this former competitor.

Infoseek offers original additional services, particularly for keeping track of current events and associated search tools. The search engine has been further improved by a new technique called Ultraseek.

Open Text is also a reference point for searching Web pages. Its performance characteristics are remarkable and merit your attention.

We could have extended this look at search engines, but this selection of the leaders will meet all of your requirements.

It is difficult to choose from these eight specialist tools for searching the Web. Sometimes you will need to overlap the results of some of them in order to optimise your search. In very specific cases, you will be able to opt for this or that one (keyword suggestions, searching in natural language, searching for chronicled sites, use of specific filters, searching for multimedia objects). You may also have your own favourites as time goes by. Learn to familiarise yourself with each of them as you carry out your investigations, take note of your favourites and install your number 1 on the home page of your browser.

And if you cannot manage to decide on this or that search engine, or if your appetite for surfing is very strong, opt for compilations of search engines and for metasearch engines.

Chapter 4

Metasearch engines

THE CONTENTS FOR THIS CHAPTER

- Metasearch engines
- Using search engine compilations

Experienced surfers all have their favourite search engine. Some swear by AltaVista, while others prefer Yahoo! or maybe UK Plus for searching UK sites. To search for information there is another solution: metasearch engines. Their principle is simple: you indicate the keywords corresponding to your request and the metasearch engine sends the request to the various search engines.

Another possibility is to use search engine groupings on the same page. This facilitates intensive searching and allows you to have a list of powerful tools at your hand.

METASEARCH ENGINES

They use the power of several search engines without you having to worry about it. They differ in terms of the choice of search engine and the evaluation of the results given to you.

▰▰▰▰ Savvy Search

http://www.savvysearch.com

Savvy Search was developed in 1995 by the University of Colorado.

A metasearch engine, Savvy Search gives you the fastest 10, 20, 30, 40 or 50 replies in searching DejaNews, Infoseek, Aliweb, WebCrawler, Excite, AltaVista, Galaxy, TribalVoice, Open Text, Lycos, Yahoo!, Yellow Pages, FTP Search95, shareware.com and Magellan.

Figure 4.1: The simple Savvy Search page

To start the search, type the keywords and click the Search button. You can refine your search by clicking the AND or PHRASE buttons according to whether your keywords should be treated as a phrase. The search itself may take some time, given that several search engines have to be contacted before the results can be displayed.

▰▰▰▰ Dogpile

http://www.dogpile.com

This powerful metasearch engine can target ten areas of the Internet for searching. Most usefully, of course, Dogpile searches the Web, presenting search results from 14 popular engines quickly and effectively. By selecting the appropriate radio button you can also search Usenet (newsgroup) archives, FTP servers, world and business news headlines or stock quotes.

Being a US site, the remaining options on Dogpile's main page are not especially useful to the UK user. You'll rarely need to search for businesses, people, maps or weather forecasts in the US. But a click on the International link at the top of the page will take you to a list of similar services covering a number of other countries. The UK is particularly well served here, with search options including Business Finder, People Finder, Local Information and Travel Guide.

Figure 4.2: The main Dogpile search page

Inference Find

http://m5.inference.com/ifind/

Inference Find is not simply a point of access to the various search engines. It can combine the searches carried out by the different search engines and return a single list to you after deleting redundancies.

Inference Find polls the best search engines on the Web: Yahoo!, Lycos, AltaVista, Infoseek, Excite and WebCrawler. Each search engine is active in parallel. If one search engine gathers 300 results at once, whereas its neighbour only gathers ten, the latter will be polled 30 more times. With Inference Find, search engines work flat out! Then all the results from the six sources are analysed. Redundancies and similar documents are eliminated in order to produce the most appropriate list of results.

It is impossible to carry out manually what Inference Find provides automatically, so why go without?

The search time can be limited, which is very often useful. Results are displayed by heading and include a list of titles only, each title giving access to the corresponding site.

Figure 4.3: Searching with Inference Find

MetaFind

http://www.metafind.com

MetaFind is also a metasearch engine. Like Inference Find, you can specify how long the search should continue before displaying its results. The results can include descriptions, and you can choose from a number of sort options such as Sort By Engine or Sort Alphabetically.

Figure 4.4: MetaFind allows timed searches and options for result-sorting

Ask Jeeves

http://www.askjeeves.com/

Put into service in June 97, this tool guides you in your search provided that the question is expressed in English. Thus you must learn to formulate your questions correctly in order to make full use of the system. If the question is complex, the program helps you to formulate your question precisely by making choices from drop-down lists.

Figure 4.5: Ask Jeeves your questions!

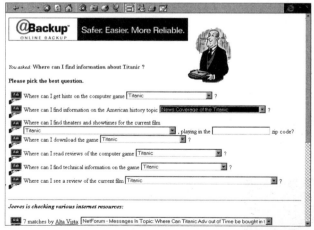

Figure 4.6: Jeeves has reformulated the question and offers several alternatives. It also indicates the initial results provided by the large search engines (AltaVista, Infoseek and Lycos)

It also offers a list of standard questions relating to your own, and it is up to you to select the most pertinent. It is very user-friendly and well worth getting to know.

Jeeves is there to help you to ask good questions. If you do not manage this, you can always activate the random question mode! Either way, you will not be short of answers.

Mamma

http://www.mamma.com

Mamma is there to take care of you and will put your request to six well-known search engines:

- AltaVista;
- Excite;
- Infoseek;
- Lycos;
- WebCrawler; and
- Yahoo!

Not forgetting DejaNews for searching newsgroups (see Chapter 6).

Select the search domain, check the appropriate option boxes (activation of summaries, searches based on a phrase, searches based on page titles) and click on Search. Mamma will do the rest for you.

Figure 4.7: Welcome to Mamma!

Meta Crawler

http://www.metacrawler.com

This is one of the veteran metasearch engines and uses several search engines to meet your requirements. Created in 1995 at Washington University, Meta Crawler was bought by the go2net company in 1997.

Also worth discovering is Mini Crawler in which a small window is opened on the screen. This is highly practical as it means that you can have a metasearch engine on standby in a corner of your Windows Office to search the Web at any moment.

Figure 4.8: Meta Crawler at your service!

Figure 4.9: The Mini Crawler window

ProFusion

http://www.designlab.ukans.edu/profusion/

Designed by Kansas University, ProFusion uses several search engines to carry out your search.

ProFusion also offers a choice of operating mode. You can select:

- the three fastest search engines; or

- the best three search engines (relevance).

This enables ProFusion to determine which are the fastest or the most relevant engines according to your requirements.

You can decide to use the nine pre-set search engines or to activate those of your choice, including:

- AltaVista, Excite, Lycos, WebCrawler and HotBot, for search engines which accept Boolean expressions;

- Infoseek, Open Text, Magellan and Yahoo! for the others.

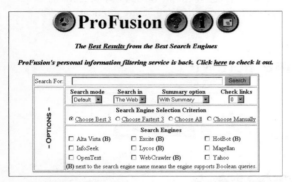

Figure 4.10: ProFusion is a very user-friendly search engine

GROUPINGS OF SEARCH ENGINES AND DIRECTORIES

These compilations of the best tools are practical for carrying out intensive searches and comparing the results of several search engines.

The Internet Sleuth

http://www.isleuth.com/

This is a fantastic tool which gives access to more than 2000 databases.

The original feature of the site is that it only catalogues sites having a search text box. The site search mechanism is displayed on Sleuth according to the requests made. Searches are carried out by subject and not by specific request. Do not ask for the recipe for duck à l'orange, since no database is concerned with this. Search for the word "recipe" instead. You will then have access to databases relating to cooking.

You thus obtain a list of hits, each entry point having its own search text box and that of the database corresponding to the initial search. At this stage you can. search for the recipe for duck à l'orange in the databases proposed. Sleuth therefore allows you to search for and find everything worth searching for ... and finding!

Six specific domains are offered:

- search engines;
- directory sites;
- news;
- business and finance;
- software; and
- Usenet (newsgroups).

The method is the same for all six categories. Enter the keywords, select the search engine or site and click on the Search button.

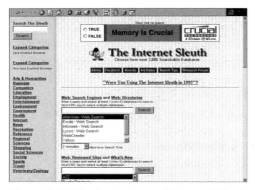

Figure 4.11: The home page of Internet Sleuth

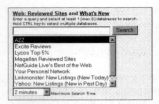

Figure 4.12: Searching in compilations of sites

On the left of the home page you can select a subject or sub-subject of interest. You can also indicate a maximum search time in order not to tie up your PC unnecessarily. By default, the value is fixed at 2 minutes.

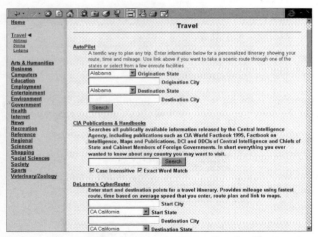

Figure 4.13: An extract from the "Travel" selection page

Aztech Cyberspace Launch Pad

http://www.aztech-cs.com/aztech/surf_central.html

This site offers a simple list of search engines and the possibility of searching directly from this page.

Figure 4.14: The choice of a metasearch engine or one of the large international search engines on the home page

▰▰▰ Directory Guide

http://www.directoryguide.com/

This service is a compilation of search tools, directories and guides – 400 of these have been referenced and are accessible by keywords.

▰▰▰ Web Taxi

http://www.webtaxi.com/

Select an international search engine, a regional search engine (by country throughout the world) or a subject, and click on the Search button. Web Taxi then displays the tool or tools found at the bottom of the screen. The rest is up to you.

> *Search engines to search for search engines!*
>
> *You can use a search engine such as Yahoo! to search for other search engines. Go directly to the following address: http://www.yahoo.com/Computers_and_Internet/Internet/ World_Wide_Web/Searching_the_Web*

Yahoo! offers the following tools in particular:

- *156 search engines!*

- *137 metasearch engines or search engine grouping pages!*

- *245 Web directories!*

Chapter 5

Directories of people and companies

THE CONTENTS FOR THIS CHAPTER

- finding people (address, telephone number, fax number, etc.)
- finding e-mail addresses
- finding companies or businesses
- consulting professional directories
- international use of the Yellow Pages

Apart from the vast source of information and services which the Web constitutes, the Internet is also a means of communication. Email is the most frequently used application, just in front of the Web. Contacting others and finding a company are the fequent needs of surfers.

Searches for people may be based on addresses, telephone numbers or e-mail addresses. There are specialist tools and services on the Web for each of these requirements. We have already seen that most large search engines (see Chapters 2 and 3) offer search functions relating to people, e-mail addresses and Yellow Pages services. This fifth Chapter gives us the opportunity to introduce some new ones and to look at the most reliable ways of tracking down people and companies.

There are millions of us around the world who use the Internet to communicate. So, if you have lost somebody's details, if you are looking for somebody who has disappeared, if you wish to make contacts abroad, this is your opportunity.

FINDING PEOPLE AND E-MAIL ADDRESSES

Some tools specialise in searching individual countries; others search the whole planet unless you choose to restrict searching to a particular country. However, in the UK (as in many other countries), finding people is by no means an exact science – there are no directories relating exclusively to UK residents. We'll begin by looking at the worldwide services most likely to lead you to an Internet user in the UK. Remember that you can actually use these services to find someone anywhere in the world by just selecting the appropriate country.

▬▬▬ In the UK

InfoSpace Email Addresses

http://in-119.infospace.com/info.jeeves/email1.htm

By default, the InfoSpace Email Addresses lookup offers a search of 'Any Country', so your first step must be to select 'United Kingdom' from the list of countries. Enter the last and first name of the person you want to find and, optionally, their home town or city. Ignore the State/Province box which relates only to American searches, and click the Find Email button.

Figure 5.1: Starting an e-mail address search at InfoSpace

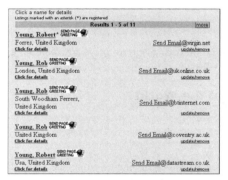

Figure 5.2: The result of searching at InfoSpace

InfoSpace has various methods of gathering its information, so the details available for each person in the results list will vary. If the person is registered with InfoSpace, you may find their address, phone number and more when you click on their name. InfoSpace doesn't display the e-mail address, but provides a link you can click to send an e-mail message.

InfoSpace Phone Numbers

http://in-105.infospace.com/info.jeeves/people.htm

Along with the e-mail address search mentioned above, InfoSpace also makes a great job of finding someone's postal address and telephone number. The form is almost identical to the e-mail addresses lookup, but there's one difference to note when completing the fields: unless the person you are searching for has an unusual name, try to enter the town or city as well as the country. If you don't, the service may return hundreds of matches!

Figure 5.3: InfoSpace is the best place to find a UK resident's address and phone number

Throughout the World

Four11

http://www.four11.com/

Four11 is still the reference tool for finding an e-mail address. All the same, the fact that your correspondent has an e-mail address does not necessarily mean that you will find it. There is no shortage of addresses! For example, it is not easy to know which is the right Bill Gates from among the 125 found! Is just one of them the boss of Microsoft? By clicking on the headers listed, you obtain more specific information and can then proceed by elimination.

The site offers many additional services, from searching for telephone numbers to sending flowers to a friend!

Also available is the directory of celebrities: actors, journalists, businessmen and women, senior officials and sportsmen and women are also listed. An ideal point of entry for contacting the "greats" of this world. You will find addresses and, additionally, e-mail addresses. Continue the search, and eventually you will find the right e-mail address for Bill Gates!

Figure 5.4: The Four11 home page

Figure 5.5: The details of a contact found, and access to e-mail

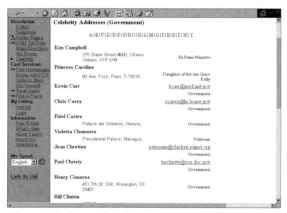

Figure 5.6: Searching for celebrities. Addresses and, additionally, e-mails addresses are listed. An extract from the Government category, letter C

WhoWhere?

http://www.whowhere.lycos.com/

WhoWhere is a famous US search service, part of the Lycos network of search engines. The service delivers highly detailed results which may include the address, phone number and hobbies of people who have registered with the service.

The WhoWhere? search engine works by approximation. The search criterion does not need to correspond to an exact entry in the index. Therefore typing or spelling errors in entering names are not too much of a problem. The search engine will in fact find overlaps of strings of two characters or more. It even allows searches using initials!

The service also offers:

- searching for addresses and telephone numbers;

- searching for companies on the Internet;

- Yellow Pages: all companies (not necessarily present on the Net). Note the Big Book service for finding companies in the United States.

Figure 5.7: The home page of the WhoWhere? Service

Figure 5.8: The presentation of replies, by family of relevance (highly, probably or possibly appropriate)

Bigfoot

http://www.bigfoot.com

The service offers:

- searching for e-mail addresses; or
- the white pages (people's addresses).

In addition to the service, Bigfoot indicates how it has obtained the information. This may be useful in making an unsuccessful search more specific.

On Bigfoot, listed users may remain anonymous if they wish. In plain language, you can find them, even send them an e-mail, but without knowing their address (this remains hidden).

Figure 5.9: The Bigfoot home page

AOL NetFind

http://www.aol.com/

America On Line, a private operator which offers a proprietary site with a gateway to the Internet, is opening up more and more on the Web. The AOL site offers a search engine using the Excite technique.

The search is carried out by name, forename and state code.

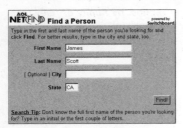

Figure 5.10: Searching for a person

Figure 5.11: The results

> ### Collections of search engines
> *If the search engines mentioned previously are not enough, or if you want to access them from the same page, use one of the All in One services for finding people and e-mail addresses, such as **http://www.albany.net/allinone/all1user.html#People**.*

Dozens of e-mail address search engines are thus accessible, in alphabetical order, from 411 Locate to Yahoo! People Search.

The same service is provided on FrontPage Express **(http: //www.thefrontpage.com/search/search.html)** *where you can select the People & Business options from the home page menu.*

 Are you looking for a telephone or fax number, an address or e-mail address **anywhere in the world?**

Use Telephone Directories on the Web **(http://www. contractjobs.com/tel/)**. *The directories of 44 countries are easily accessible. The service is quite austere, but highly efficient.*

For the United States, the list of directories offered is impressive. First of all the white pages: AnyWho, Switchboard United States White Pages, Yahoo! People Search, Bigfoot, Infospace, Lookup United States, Database America, WhoWhere? and pc411.

Next, the yellow pages: BigBook, Zip2, US West, Big Yellow ninex Interactive Yellow Pages, Yellow pages online, Onvillage Yellow Pages, GTE Superpages, YellowNet and True Yellow.

To all this must be added a fax directory and directories for freephone numbers (0800).

TRADE MARKS AND COMPANIES

Directories of trade marks and companies are legion on the Net. Here is a useful selection. You will need to distinguish between tools which only list companies present on the Web, and the others, of more general scope.

▬▬ The Patent Office – Database Search

http://dips.patent.gov.uk

The Patent Office deals with UK trademark and patent applications. While the main site at **http://www.patent.gov.uk** provides useful background information, the Database Search site allows you to search patent databases in the UK, Europe and WorldWide.

Europages

http://www.europages.com/

This European directory lists 150,000 companies spread over 25 countries in Europe.

Figure 5.12: Finding companies using Europages

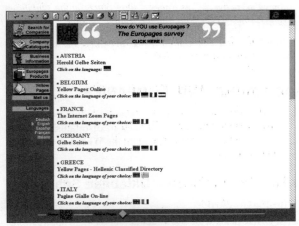

Figure 5.13: Finding directories of companies by country

You can search in three ways:

- in plain text mode;
- by business; or
- by company name.

Note the Yellow Pages button **(http://www.europages.com/yp-en.html)**
which allows you to search through 17 European countries. Select the
country and one of the languages offered (by clicking on the flag).

You then access a directory of companies for the country selected.
Europages thus gives you easy access to the best company
directories throughout Europe.

The Biz

http://www.thebiz.co.uk

The Biz provides one of the simplest ways to locate a UK company,
whether the company you are searching for is present on the Internet
or not. Using a directory structure similar to the UK Plus and Yahoo!
web search engines, you work through categories and sub-categories
until you find the company you are looking for. If that company
has a web site, clicking the link will take you to the company's
site. If not, the link will take you to a page giving contact details
for the company.

After selecting a primary category from The Biz's front page, you
can choose between four methods of displaying company links:

- All Entries: View all the companies and sub-categories in
  the category you selected;
- All entries by County: Display only companies from your
  chosen county in the current category;
- All Entries by Region: View companies from a particular
  region of the UK (such as Scotland or West Country);
- All Entries by Name: Click a letter to display only
  companies whose name begins with that letter.

Figure 5.14: The General category of The Biz's company listings, using All Entries view

If you're not keen on delving through categories to find a company, The Biz also offers the traditional keywords search from its page at **http://www.thebiz.co.uk/search.htm**. Your keywords may include the name of the company or organisation you want to find, or words you would expect to find in a description of the company.

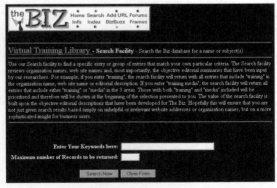

Figure 5.15: Find companies and organisations at The Biz using a simple keyword search

CompaniesOnline

http://www.companiesonline.com

This is yet another service that forms a part of the huge Lycos network, and can provide detailed information about a large number of US companies. Along with contact details this information may include annual turnover, number of employees, the latest stock quote and much more.

Like The Biz, CompaniesOnline lets you choose between running a traditional search or 'drilling down' through category listings.

Figure 5.16: Search for US companies by name, industry or directory category

LookHere Web Directory

http://www.lookhere.co.uk

If you know that the UK company you're looking for has a presence on the Web, LookHere may be able to take you to its site. LookHere's categories are organised by initial letter, so clicking 'C' on the home page will lead to a list of categories that includes Cars, Cinema and Construction. Click a category to see a page of links to companies in that category, accompanied by brief descriptions.

THE YELLOW PAGES

The expression "Yellow Pages" is widely used on the Web. These search sites allow you to find details of businesses and companies on the Internet.

With the formidable development of companies on the Web, there are plenty of Yellow Pages services. The large search engines have them, as we have already seen.

▬▬ Yell

http://www.yell.co.uk

Yell is the online incarnation of the UK Yellow Pages telephone directory. Its great benefit is that it covers the entire country rather than limiting you to searching in your own telephone area. You can choose between two methods of searching. Either select a business type (such as Painters and Decorators or Airlines) or search for a specific company by name. If you leave the Location box blank, Yell will return search results for the whole country.

Figure 5.17: Searching for businesses and companies at Yell

Along with the name, address and phone number of businesses matching your search criteria, small hyperlinked icons may be included to indicate that the business has its own web site, or provides a map or additional information.

---

yellow pages search
━━━━━━━━━━━━━━**results**

Your request was for HOTELS & INNS in UNITED KINGDOM.
Yellow Pages Search has the following matches:

Sherborne Lodge Hotel	Torrs Park Ilfracombe Devon EX34 8AY **Tel: 01271 862297** **Fax: 01271 865520**
Sherbourne Hotel The	Sutton St Tenby Dyfed SA70 7DX **Tel: 01834 843980** **Fax: 01834 843980**
Ship & Bell Hotel	6, London Rd Horndean Waterlooville Hampshire PO8 0BZ **Tel: 01705 592107**
Skelwith Bridge Hotel	Skelwith Bridge Ambleside Cumbria LA22 9NJ **Tel: 015394 32115** **Fax: 015394 34254**
Smugglers Inn	14, Harbour Place Burntisland Fife KY3 9DP **Tel: 01592 873882** **Fax: 01592 873882**

Figure 5.18: Yell's search results include clickable icon links to extra information

▬▬▬ Scoot

http://www.scoot.co.uk

Scoot provides a similar set of services to Yell, allowing you to search for a type of business or a company name, selected from links to the right of the search form. A useful extra is the Product Finder page, which can help you find businesses that provide the particular product or service you're looking for.

Figure 5.19: Search for a product using Scoot's Product Finder directory

---

▬▬▬ **Kompass**

http://www.kompass.com

This is a monster which lists 1.5 million companies throughout the world, spread over 61 countries!

There are two search modes:

- product search; or
- company search.

Figure 5.20: Welcome to the search for companies throughout the world

Figure 5.21: One stage in the search – selecting the country

Let yourself be guided by the search procedures the first time. Then use Help to optimise your searches and understand all the fine points of the system, as it is easy to get bogged down.

World Wide Web Yellow pages

http://www.mcp.com/newriders/wwwyp/index.html

Provided by Macmillan Publishing, this service offers a search for companies by keywords or business.

Figure 5.22: Searching the WWW Yellow pages

Figure 5.23: An example of a result provided by WWW Yellow Pages

Infospace

http://www.infospace.com

Finding companies is the heart of this service. But you can also find white pages, and search for online purchasing sites, small ads, maps, etc.

Infospace offers several modes of operation for finding companies:

- by category;
- by name;
- by address, even if only approximate;
- by telephone or fax number; or
- by map (USA, Canada and London).

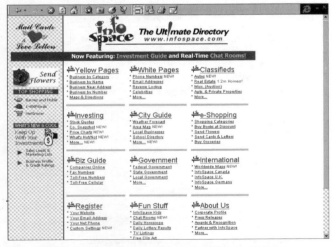

Figure 5.24: All the search tools accessible on Infospace

▬▬ LinkStar

http://www.linkstar.com/

Here again is a service which allows you to find companies throughout the world. It offers multicriteria searching in a database of 320,000 entries. The zone for entering keywords may be completed with various options:

- company name;

- contact name;

- address;

- type of business (drop-down list); or

- country (drop-down list).

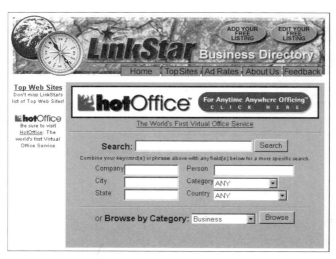

Figure 5.25: The LinkStar search page

Chapter 6

Searching with Usenet and FTP

THE CONTENTS FOR THIS CHAPTER

- Finding newsgroups
- Finding downloading sites
- Finding shareware

With the huge success of the Web, we have come to confuse the Web and the Internet. In fact, the Internet has a set of older resources such as:

- Usenet: for newsgroups or discussion groups;
- FTP (*File Transfer Protocol*): downloading sites.

These resources are increasingly accessible on the Web owing to their user-friendliness.

NEWSGROUPS

Your Internet connection gives you access to:

- the Web;
- e-mail; and
- newsgroups.

Accessing newsgroups

Newsgroups, at least those offered by your Internet service provider, are accessible with your e-mail software. Outlook Express, for those who have Internet Explorer 4.0 or higher, will thus be used to find newsgroups, consult them and participate in them.

Here we give a few paths for accessing newsgroups with this software. The basic principles are the same with any other e-mail software and newsgroups.

A newsgroup corresponds to a subject of discussion: HTML language, science fiction, politics, cyberculture, etc. Hundreds of newsgroups are open to you. No point hiding your face in horror: many newsgroups have a strong sexual connotation. All sites with the name **alt.sex** are of a pornographic nature, but these are not the only ones. So the more serious subjects, such as object-oriented programming, can exist side-by-side with the naughtiest subjects. It is up to you to sort them out.

If we ignore the broadcasting of pornographic photographs and messages inviting you to connect with Web sites of the same nature, newsgroups are a good way of participating in life in cyberspace, communicating with others and exchanging information on common interests. Again, you need to find the right addresses!

Newsgroup addresses take the form *domain.subject.sub-subject*; e.g.: **alt.music.pop.**

Domains are represented by codes: **alt** for alternative, **comp** for newsgroups connected with computers, **rec** for recreational or **talk** for chatting and exchanging ideas.

Figure 6.1: Finding newsgroups using Outlook Express

The All tab in the above figure provides a list of all the newsgroups accessible. The Subscribe tab contains a list of the newsgroups to which you have decided to subscribe.

Chat groups operate off-line. You read the messages posted, but you are not in direct communication with their authors. Likewise, when you post a message, this will be read by whoever wants to whenever they want. Interactive, or real time chatting, requires the IRC (or chat) technique, which we will deal with later.

You may have access to almost 50,000 different newsgroups, and many of these will be 'worldwide' groups with contributors from anywhere in the world (although the dominant language is English, as elsewhere on the Internet). For some types of information search this wide range of contributors means that you should be able to find what you want more quickly.

There may be times, however, when only UK-specific information will do: sifting through a worldwide newsgroup for UK legal advice would be a painful process! Fortunately there are a number of UK newsgroups covering a large number of subjects. These are easy to find since their names all begin with 'uk'.

Hundreds of newsgroups, sorted by alphabetical order, are therefore directly accessible using your e-mail software. By typing a few letters in the data entry zone, you access specific newsgroups faster. For example, type **uk** to go to the list of UK-specific newsgroups. Select a newsgroup and connect in order to consult it.

You will then learn:

- how to subscribe to a newsgroup;
- how to read messages off-line;
- how to participate in newsgroups by sending messages; and
- how to find, organise and sort messages.

In fact, the enormous number of messages of some newsgroups makes them difficult to consult. However, search functions make it easy to access the information you want. You can thus carry out searches by indicating:

- the sender's name;
- the object (in full or in part), a keyword suffices; or
- validity dates (posted before, or after).

FINDING NEWSGROUPS ON THE WEB

Once again, the Web is there to help you to find newsgroups. Some search engines offer this service, such as DéjàNews, of which it is a speciality.

▬▬▬ DéjàNews

http://www.dejanews.com/

DéjàNews offers the following functions:

- **Quick search**. Traditional searching by keywords;

- **Advanced search**. Use of logical operators (OR/AND);

- **Search filters**. Creation of filters by group, author, subject and date;

- **Interest centres**. Search for groups corresponding to a given subject;

- **Newsgroups**. Search by group name and navigation through the newsgroups hierarchy – you go to a branch and start a specific search where you want.

When you start a simple search by keywords, you can specify a search domain from the following options (drop-down list):

- complete: all newsgroups;

- standard;

- adult: erotic newsgroups;

- job: newsgroups devoted to employment.

Standard mode is offered by default.

The results of a search provide a list including:

- date;

- subject;

- newsgroup name; and

- author.

Each element of the list is a link on which you just need to click in order to obtain:

- the detailed content of the message;

- additional Web access if a link is referenced in the body of the message.

From a given message you have six commands:

- Display previous message (list of results);
- Display next;
- Return to list;
- Display author's profile; and
- Reply to message (post).

Figure 6.2: The DéjàNews logo

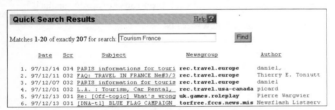

Figure 6.3: The result of a quick search

Figure 6.4: The Power Search method

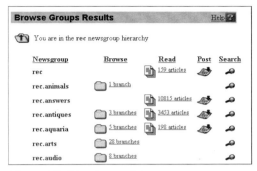

Figure 6.5: Direct search by newsgroup name

Save your search filters

The use of filters provides the user with an effective way of finding newsgroups. It is thus possible to specify ranges of dates, to limit oneself to certain subjects or to search for the messages of a particular author. Once the filter has been created, it is saved on the DejaNews server, allowing it to be used again later.

FINDING DOWNLOADING SITES

Downloading sites are used to import all kinds of things onto your PC:

- programs, utilities and plug-ins (software adding functions to your Internet browser);

- shareware;

- drawings, images, photographs;

- sound files;

- videos;

- text documents.

The advantage of downloading software from a Web site rather than an FTP site is that it is more simple. Instead of going through directories, you let yourself be guided. And that avoids the need to use several different softwares.

FTP Search

http://ftpsearch.ntnu.no/

This Norwegian site offers the possibility of searching for downloading sites. Be warned, this service only works using file names and path names (directories). No searching by keywords or subject with this tool. This supposes that you already have information making it possible to locate and reference the object to be imported.

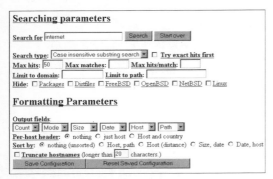

Figure 6.6: The ftp site search screen

The service allows you to format hits and save parameters. It is possible to declare the maximum number of hits (max hits), eliminate certain types of file (hide) or limit the search to a specific domain.

FINDING SHAREWARE

Shareware are programs which offer a free trial. Downloaded from the Web, they meet many requirements. But you still have to know how to find and sort the good, the very good even, from the bad!

Shareware.com

http://www.shareware.com

Shareware constitutes a valuable tool which you can procure directly from the Net. Whatever your need, there is surely software which can satisfy it. And shareware is to be found almost everywhere on the Internet. If only there was a site offering all existing shareware! Where you could get to know all the versions and be sure of downloading the latest. Where you could also find products for Mac, PC, those that run under DOS, Windows 3.1 or 95 or even NT. One can dream …. There would also need to be a short introduction to each software in order to know what it is used for and what its limits are. But wait, this service exists! It is called shareware.com.

Three different search modes are offered from the home page screen:

- quick search;
- simple search; or
- power search.

Whichever mode used, you will have to enter and specify a number of criteria.

The results of a search reveal:

- the name of the file (the software) to be downloaded;
- a brief description of the program;
- the storage directory (category of program);
- the date of the file; and
- the file size.

Figure 6.7: Finding software using Shareware.com

Figure 6.8: The result of a quick search using the keyword "gif"

One good thing: the most recent programs appear at the top of the list ("new" label). It is up to you to choose!

Shareware – how to use it
Shareware is not free software. It is a program which you can try out free of charge. In plain language, you can download it freely and install it on your hard disk. If the program suits you, and you wish to keep it, you must pay the author. The price of the license and the programmer's details are generally mentioned in a "readme.txt" file or are accessible when you start the product. Often, it is possible to register and pay for the product on-line. It is good to pay for the licence, it is one of the rules of the game. Moreover, you will then be able to benefit from upgrades or receive more detailed documentation. On the other hand, if the program does not meet your requirements, just delete it from your hard disk.

Downloading time depends in theory on the size of the file and the speed of your modem. In practice, this time also varies depending on how congested the site and network are.

To download shareware, you must:

1. Copy the file to a directory on your hard disk.

2. Decompress the file.

3. Install the program (run the file and follow the on-screen instructions).

You will therefore need to have a decompression program. Most shareware supplied for PC is compressed in the Zip format. You must use the Winzip program to decompress it. So you need to download Winzip, which is also shareware. Fortunately, it will not be compressed and you will be able to make the best use of it!

On a Macintosh you will be able to download Stuffit Expander, for example. But be warned, you may need Stuffit to import Stuffit!

Filez

http://www.filez.com/

The Filez site offers 75 million files to download at the click of a mouse. Whether you are looking for software for a PC or a Macintosh, you can use Filez. Once identified, the software that you want can be downloaded directly. FTP sites (more than 5,000) mainly contain shareware, but Filez also gives access to several thousands of different commercial softwares (the price varies depending on the vendor chosen). It is up to you to hunt down a bargain!

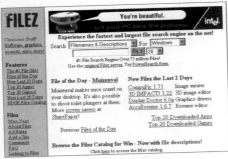

Figure 6.9: Downloading shareware and commercial software

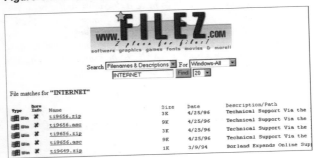

Figure 6.10: The results of a search

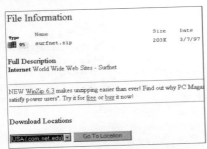

Figure 6.11: A program description and the choice of downloading location

Chapter 7

Multimedia searches

The Contents for this Chapter

- Finding images and sounds
- Connecting to photo agencies
- Finding maps

Finding Images

It is not difficult to find images and, more generally, multimedia objects on the Web. Image and multimedia files are easily identified by their extension. It is easy to sort between:

- images in Gif format;

- images in JPEG format;

- sounds (wav…);

- videos; or

- animation.

Gif and JPEG are in fact the two formats most used on the Web. The Gif format records an image in 256 colours. Developed by Compuserve, it is a highly compact format which is well suited to transfers on slow-speed networks such as our good old switched telephone network. To obtain higher quality, use the compressed JPEG format (*Joint Picture Expert Group*). This format is better suited to transmitting high-quality photographic documents. But, in order not to slow down the loading of pages, images are sometimes presented in the form of small images, called 'thumbnails'. You can open the image on a new Web page in its true size and with the best possible quality just by clicking on it.

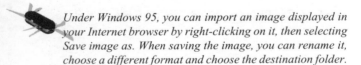

Under Windows 95, you can import an image displayed in your Internet browser by right-clicking on it, then selecting Save image as. When saving the image, you can rename it, choose a different format and choose the destination folder.

Image search engines

Image search engines allow you to scan the Web looking for illustrations, photos or rare documents. Be careful; if the documents retrieved are not free of charge, you will not be able to publish them yourself, unless you pay the author. But for your personal use, you can amuse yourself by collecting everything which interests you.

Webseek

http://www.ctr.columbia.edu/webseek/

This is an extraordinary tool which we owe to Columbia University in the United States. All image hunters should connect with this site in order to try it out. Webseek offers browsing by menu in addition to searching by criterion. Insofar as images are concerned, the result of your request is presented in the form of a page of thumbnails (see Figure 7.3), and the size of the original images is indicated. You can go to the site where the chosen image is located with just one click and open the image in its original size.

The system allows you to find videos, black and white photos or colour photos. To do this, just check the appropriate box in the keyword text box.

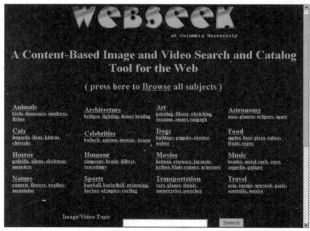

Figure 7.1: The home page with searching by subject

The result is therefore presented in the form of an electronic contact sheet: each corresponding document is displayed in the form of a small image. Then you just click on the image to access a document. The size of the actual image is displayed above the small image.

The service even allows you to find photos with similarities in terms of colours.

Figure 7.2: Searching by keyword and media type

Figure 7.3: The result of a search

Figure 7.4: The His command (histogram) allows you to search for similar shades

Figure 7.5: Searching by subject: here: Arts – Painting, then Matisse

Lycos Image

http://www.lycos.co.uk/search/options.html

You discovered the general-purpose search engine Lycos during Chapter 2, but the speciality of Lycos Image is, as its name suggests, that it finds images. To do this it uses a specialised analysis technique which is truly sophisticated. For example, the search engine allows you to find tourist images or photographs of well-known personalities.

Finding sounds on the Web

Lycos also allows you to find sound files. Take care to check the file formats (wav, midi, au or ra) and to ensure that you have the right software to listen to them. A number of sound documents broadcast are coded in ra format (for Real Audio) and can be processed by the Real Player software of the Progressive Networks company. This technique also allows sound to be broadcast live (live radio on the Web). Note that the Real Player program is included in Internet Explorer 4 and 5.

Yahoo! Image Surfer

http://isurf.yahoo.com/

You have already discovered Yahoo! UK & Ireland, then the international version of Yahoo!, and now here is Yahoo! Image Surfer. This tools allows searching by keywords or from among the 6 categories offered (arts, leisure, "people", games, science and transport).

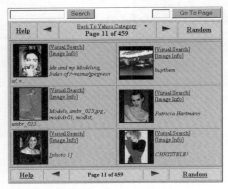

Figure 7.6: Searching by category; here a page on the top models

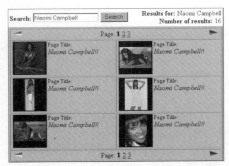

Figure 7.7: The request has been specified with the keyword "Naomi Campbell"

Once you have selected the category, a panel appears at random. You can specify the object of your search from the text box at the top of the screen.

Here again is a very efficient tool which image hunters should save in their favourites.

Finding erotic sites

We cannot allude to searching for images without mentioning eroticism and pornography. If there is one domain where photographs are used in profusion, it is this one. And there is no shortage of such material on the Internet.

*Do we need a search engine for this? Why not? Those who want to get down to the nitty gritty will be able to connect with **http://www.sexhunt.com/**.*

Photo agencies

Photo agencies on the Web constitute a good point of entry for browsing through images. These are professional sites offering the best negatives of the best photographers or offering high quality images for sale. In order to forearm themselves against abuse, images are displayed in low resolution and with limited colour palettes. In any case, it is not possible to display photographic quality images on the Web, as the pass-band of networks is not suited to this. However, these documents may be offered for downloading.

Sygma Agency

http://www.sygma.fr/

The Sygma agency is one of the largest photo agencies in the world. The site presents a stock of almost 500,000 images which you will in future be able to import in high definition for a fee. This service, Sygma Direct, is in the pipeline.

The results of a search by keyword are presented in the form of a contact sheet. By clicking on one of the images, you open the document in a separate window, and at the same time you obtain the photo's references.

You will also be able to benefit from online reports (Stories heading). News, People, Magazine and Illustration are the four sections from which you will be able to access recent photos.

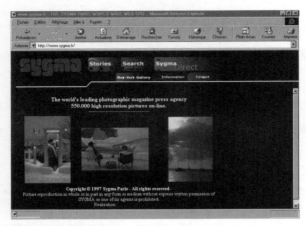

Figure 7.8: Welcome to the Sygma agency

Figure 7.9: The result of a search on the word "portrait"

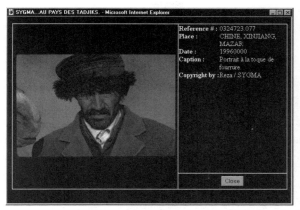

Figure 7.10: A photo is presented on request in a separate window

Giraudon Agency

http://www.giraudon-photo.fr/french/seek_cnt.htm

The Giraudon photographic agency has a unique base of 200,000 negatives and 70,000 ektachromes combining photographs in various categories including Decorative Arts, Celebrations, Mythology, The East, and Daily Life.

Searches are not carried out automatically on the Web. You must complete a form and indicate your wishes. It is explained that the document base is particularly rich. If you are interested in impressionism, more than 1,000 references are in stock. You can find almost 500 documents concerning Napoleon, for example. And if laughter is the subject of your search, almost 100 documents from Gargantua to the Laughing Cow will be offered!

Finding sounds or videos

*For sound you can use the Lycos service at: **http://www.lycos.co.uk/search/options.html**. As we also pointed out in Chapter 3, Hot Bot is an excellent tool for* *finding sound, video and even Shockwave animations. These animations use a Macromedia technique (http:/www.macromedia.com) and you must first download the "player" from this publisher's site in order to be able to view them on your PC.*

FINDING MAPS

You have been able to see that some general search engines offer map searching. In addition there are sites which specialise in this type of search. If search engines are handy for getting you around cyberspace, maps and plans will be of use to you in getting back to earth!

▬▬▬ Maps Database

http:/www.internets.com/smaps.htm

This is a maps search engine which also offers a very full list of sites where maps can also be found, with searching by category.

▬▬▬ Xerox Map Viewer

http://pubweb.parc.xerox.com/map/color=1/features=alltypes

This software, developed by Xerox in 1993, allows you to discover the world interactively. Set the parameters of the visual representation software, click on the globe and discover the world. Latitude and longitude will no longer hold any secrets for you.

▬▬▬ The Perry-Castenada Library Maps Collection

http://www.lib.utexas.edu/Libs/PCL/Map_collection/ Map_collection.html

230,000 maps are listed and accessible from this address. There is something of everything here, and the map lover will be in his element. Maps are classified by continent and region. In addition, impressive lists are offered. On the menu are geographical maps obviously, but also historical, political, economic maps, etc.

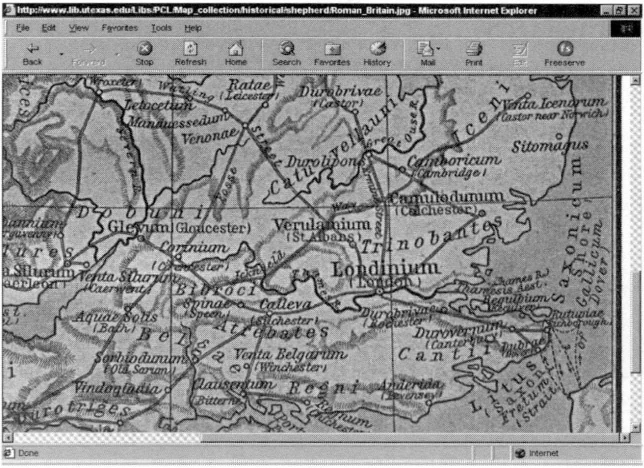

Figure 7.11: As an example, a map of Roman Britain from the historical atlas

Lycos – MapBlast

http://www.proximus.com/lycos/

This is a remarkable service which displays the corresponding map from an address (street and town), with a cross indicating the street number requested. You can also title the map in order to personalise it and possibly send it to a third party. Once the map has been displayed, it is possible to zoom in to obtain greater detail and consult all the surrounding services (hotels, restaurants, banks, etc.).

The Zoom In command allows you to obtain greater detail, while Zoom Out gives a more general overview of the place (just click on the graduated scale at the right of the map).

Just one regret: this service only works in the United States. To make up for this, Lycos offers the City Guide service which you can reach by clicking the 'City Guide' link or by going to **http://cityguide.lycos.com**.

From a map of the world, you click on a continent, then, from the new map, on a country from the list displayed. From the country, you then access the map in question and a list of towns. For example, more than 60 English counties and cities are thus accessible, and there are thousands of towns and cities around the world which you can discover with a few clicks of the mouse. For each town or city, City Guide offers you a summary and a list of links used (culture, recreation, practical, etc.).

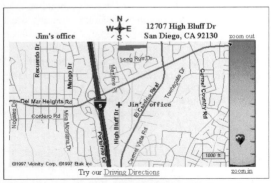

Figure 7.12: Entering an address with MapBlast on Lycos

Figure 7.13: A personalised map is displayed

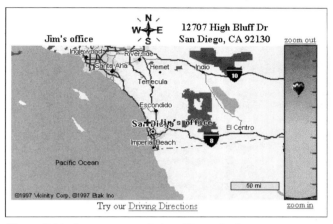

Figure 7.14: A more general view (Zoom Out)

UK Street Map Page

http://www.streetmap.co.uk

This site currently only has street maps for Greater London, but it can also provide road atlas maps for the whole of mainland Britain. A simple search box lets you search by entering a London street, a postcode, a town or city, or (if you prefer to do things the difficult way!) an Ordnance Survey grid reference or latitude and longitude coordinates.

By clicking on an area of the map you can centre the map on that grid square. The Zoom In/Zoom Out links let you view the centre of the current map in greater detail.

Use the URL http://www.streetmap.co.uk/streetmap.dll?grid2map?X=528250&Y=181250 to link to this map
And click here to find out other ways to improve the link or link by post code
[Zoom In] [Zoom Out]

© www.streetmap.co.uk

Your Street is somewhere in the centre grid square.

Figure 7.15: The result of a search for a London Street

Chapter 8

Practical searches

THE CONTENTS FOR THIS CHAPTER

- From knowledge ... to consumption
- Ferreting in virtual libraries
- Finding shops and shopping centres
- Buying books

From consulting a document to buying a product on the Web, everything starts with searching. The Web offers you virtual libraries and shops so that you can satisfy your thirst for knowledge or your hunger for consumption. To put it plainly, you are can enrich yourself (intellectually) or spend without counting the cost!

FERRETING IN VIRTUAL LIBRARIES

The hypertext technique, which was the main factor in designing the Web, was set up in order to easily exploit any kind of text document and to connect them by means of links. The written word occupies a place of honour on the Web, and you can discover different types of document by browsing through the virtual libraries available.

Library of Congress

http://lcweb.loc.gov/

The Library of Congress contains more than 110 million documents in more than 400 languages. Most of the information in the library's catalogue is available online.

From the home page, the first search zone, American Memory, gives access to the following:

- text documents;
- photographic resources;
- the first films;
- sound documents.

From the Exhibition section you access the great exhibitions. Among these exhibitions you'll be able to read transcriptions of the Gettysburg Address in 29 languages, find out how the American Declaration of Independence was drafted and read the first 'rough draft', or take a detour to the Vatican library with its own range of exhibits. Also discover the revelations of the Russian archives, or even, in another register, the discovery of America by Christopher Colombus in 1492. Or view on your PC the earliest map of California dating from 1562!

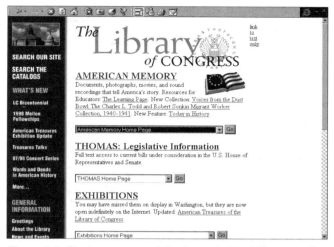

Figure 8.1: The home page of the largest virtual library in the world

The library's services are detailed in Library Services. Here you will be able to access the virtual reading rooms. 20 rooms await you with research domains as varied as history, science, the arts, rare books or the cinema. For example, connect to the Early Motion Pictures page (1897–1916) to carry out research into early cinema productions.

It is in the Research section that you will find the search tools (Catalogue page / Research and Reference command).

The Catalogue page gives access to all search modes:

- by word;
- by reference;
- by Boolean expression.

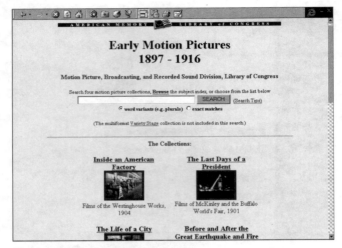

Figure 8.2: Searching for early films

About a dozen additional search tools add to the basic mechanisms.

Created in 1800, the Library of the American Congress will be celebrating its bicentenary in the year 2000! Preparations are in progress and are revealed on the Web site. Do not wait for the end of the millennium to visit this extraordinary document base.

The search engine of the Library of Congress

Named LOCIS it contains more than 27 million documents stored in different databases. Equipped with a real turbo, the library's catalogue contains not just its own documents, but also other documents found in other libraries or research institutes. Amongst these documents, for example, you will find books, magazines, maps, music scores, films, posters, photographs or manuscripts. One file allows you to contact more than 13,000 research bodies throughout the world.

The British Library

http://www.bl.uk

The hub of the British library is Blaise, the British Library Automated Information Service, a collection of searchable databases containing over 19 million records which includes:

- the British Library Catalogue, which offers access to books published all over the world from 1450 to 1975;

- Whitaker, a weekly-updated database of British books published since 1965;

- Eighteenth Century Short Title Catalogue, records of all types of eighteenth century printed materials;

- the Stationery Office, a database of government and official publications since 1976.

The full Blaise service is available by subscription, and the British Library web site does little more than give a taste of what's available. However, the main database collections can be searched online without subscription using the OPAC 97 service.

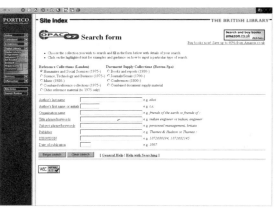

Figure 8.3: Searching the British Library

▬▬▬ Internet Public Library

http://www.ipl.org/ref/

This is a virtual library which offers you hundreds of thousands of works and documents online. Choose a domain or a discipline from the home image, then refine your search until you find the exact subject which interests you. This service will display many online references and as many links with other sites containing information relating to your research.

Also click on the home image desk to ask a specific question. It can be very entertaining to consult the list of questions asked (with the answers). All kinds of things are there – how much is a 1983 Ford Fiesta worth? Why is the year 2000 a problem for computers? How much money did Yahoo! earn last year? Why is the sky blue? (Good question!)

This is another address which you should not miss.

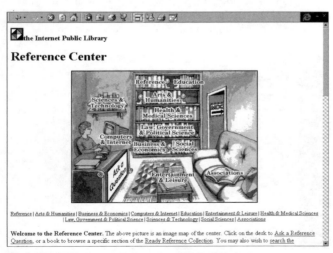

Figure 8.4: Welcome to Internet Public Library

▬▬ Athena

http://un2sg1.unige.ch/athena/html/athome.html

This virtual library created in Switzerland is the work of Pierre
Perroud, lecturer in philosophy at the Collège Voltaire in Geneva.
More than 3,700 literary texts are digitised, with the bonus of links
with other consultation or downloading sites (**http://
un2sg1.unige.ch/athena/html/booksite.html**).

Figure 8.5: Consult digitised books from Athena

*The Gutenberg Project is the most ambitious project of the
end of the millennium. Its objective is to digitise works and
issue them via the Internet. Therefore the service is
developing day by day, with new entries being added all
the time. Dante's La Comédie is the 1000th work thus
digitised. It is accessible at the following addresses: **http:/
/www.promo.net/pg, http://www.gutenberg.net** or **ftp://
ftp.prairienet.org/pub/providers/gutenberg** (downloading
site).*

WINDOW SHOPPING ON THE WEB!

After ferreting through the bookshelves of libraries, you are going to be able to browse through the shelves and racks of the Web's virtual shops. And you will always be able to satisfy your hunger for knowledge by buying books!

While online shopping has been slow to develop in the UK, the United States are literally overflowing with this type of opportunity. And there is no shortage of shopping centres and different kinds of shop. Here too you need to find your feet and select the best services. The polemic surrounding payment guaranteed by bank card was inevitable, but one has to admit that it is no more risky today to buy online than to pay by visa card at an actual shop!

Our browsers now integrate the necessary functions for protected payment, so the risk of an ill-willed third party intercepting and using your card number, which is itself encrypted, is very slight.

Therefore you can buy (or rent) just about anything on the Net:

- products: books, clothing, cars;
- services: holiday lets;
- intangible products: electronic magazines, software, stock exchange transactions, etc.

For us, in the UK, we need to check the cost and time taken to receive goods from the United States.

Buying books

Books are some of the products best represented and most sold on the Net (other than computer products: hardware and software). You will have as much fun visiting these sites as ferreting in libraries. And as a bonus, you can order the works of your choice. Setting out to find an author, a forgotten title or a rare edition is a treat!

Waterstone's Online

http://www.waterstones.co.uk

One of the biggest names in High Street book sales is also on the Web, letting you browse titles by subject, read descriptions and buy online. You can carry out a simple search by typing the name of the author, book title or ISBN into the search field that appears on almost every page of the site. If this simple search doesn't turn up trumps, click the Advanced Search link. Advanced searches are made using any combination of:

- the author's name;
- the book's title;
- the subject;
- the ISBN;
- words descriptive of the book or subject;
- the publisher.

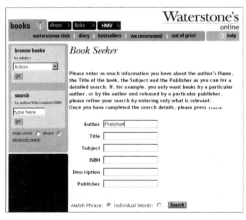

Figure 8.6: An 'Advanced Search' at Waterstone's

When you find the book you want, you can order it instantly by checking its Order box and clicking the Update order button at the foot of the page.

No	Title	Price	Availability	Order?	Quantity
1	Science of Discworld Terry Pratchett, *Hardback*	£14.99	Standard	☐	1
2	Johnny And the Dead Terry Pratchett, *Paperback*	£5.50	Standard	☐	1
3	Pyramids Terry Pratchett, *Hardback*	£8.95	Standard	☐	1
4	Jingo Terry Pratchett, *Hardback*	£5.99	Standard	☐	1
5	Johnny and the Dead Terry Pratchett, *Hardback*	£9.99	Standard	☐	1
6	Carpet People Terry Pratchett, *Hardback*	£11.99	Standard	☐	1
7	Only You Can Save Mankind Terry Pratchett, *Hardback*	£9.99	Standard	☐	1
8	Mort Terry Pratchett, *Hardback*	£14.99	Standard	☐	1
9	Dark Side of the Sun Terry Pratchett, *Hardback*	£14.99	Standard	☐	1
10	Johnny and the Dead Terry Pratchett, *Hardback*	£12.99	Rapid	☐	1

Page 1 of 10.   Next ->

You found 160 books.
Displaying the closest 100 matches.

Figure 8.7: The results of a search for fantasy writer Terry Pratchett

BOL

http://www.uk.bol.com

BOL is a new but rapidly expanding online bookstore with French, German and Dutch versions as well as this UK store. You can take your time and browse through any of 18 categories of features and reviews or carry out a quick search by book title, author, keywords or ISBN using the pull-down list and text box included at the top of every page.

BOL uses an easy-to-follow shopping basket metaphor for online purchase. Click the Add To Basket icon for any book you want to order. You can then choose to continue browsing or go to the checkout to provide delivery and payment details.

Figure 8.8: Run a quick search or browse by category at BOL

iBS: The Internet Bookshop

http://www.bookshop.co.uk

iBS is part of the well-known High Street store, WH Smith, and has one of the countries largest online bookstores with over 1.4 million UK and US titles. The usual options to browse by category or run a quick search are there, along with a Full Search option. This comprehensive search includes the usual search criteria (title, author, ISBN, publisher and series title), but also allows you to choose how the results should be presented, with options such as:

- Price (cheapest first);

- Alphabetically by Title;

- Publication date (most recent first).

Figure 8.9: The Full Search page at iBS

Amazon.com

http://www.amazon.com

With more than 2.5 million books online, Amazon has become the indisputable reference point for anybody wanting to buy a book on the Web. Even the rarest editions or books which are out of print are sold, although the service requires 6 months to track down a rare work. When the book is found, a price proposal will be sent by e-mail. You can then decide whether or not to buy the work in question.

Research is by keyword or subject. Readers can give their opinion on a work, just like authors and publishers, which is an excellent way of enriching the information base.

Amazon.com offers numerous services, some of which are as commonplace as choosing the wrapping paper for a gift. But it is these little extras which end up making a site a success.

Figure 8.10: Welcome to Amazon, the largest virtual bookshop on the Web

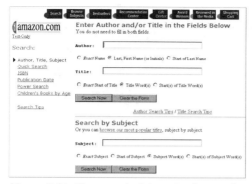

Figure 8.11: The search data entry form using Amazon.com

Barnes and Noble

http://www.barnesandnoble.com/

With more than one million references, the virtual shop window of
the famous publisher is highly attractive.

*Figure 8.12: The home page of the Barnes and Noble virtual
bookshop*

Figure 8.13: The books search engine

Figure 8.14: The result of a search

As always, searching is by category, title or author's name, or even by keyword. Just check the box of a work to add it to your virtual trolley before entering ordering mode.

Shops On The Net

http://www.shopsonthenet.co.uk

Hundreds of virtual shops at the click of a mouse-button. Browse through 30 categories of shops and services, run a quick keyword search, or click the Power Search link to search by product, shop name, country, description, and a number of other optional criteria.

Figure 8.15: The home page of the Shops On The Net shopping directory

UK Shopping City

http://www.ukshops.co.uk/enter.shtml

The UK Shopping City is a large virtual shopping centre with a number of areas including Retail, Travel and Property. You can browse by selecting an area and touring the different levels of the shopping centre where you'll find some of the biggest High Street names including Marks and Spencer, Interflora, Argos, Comet, and Dixons.

Figure 8.16: Level 1 of the UK Shopping City's Retail area

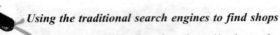

Using the traditional search engines to find shops

*For example, Web Crawler (**http://webcrawler.com/ Shopping/**)offers a search of online shopping sites. You thus have hundreds of virtual shops at the click of a mouse, classified by product type. You can find the great classics, Barnes & Noble for books, Disney Store, CD Now, etc.*

Chapter 9

"Surfing" guides

THE CONTENTS FOR THIS CHAPTER

- How to surf
- Discovering Web site compilations and guides
- Visiting top sites
- Browsing the entire world
- Searching in any language

HOW TO SURF

Surfing is not a search method in the strict sense, and yet surf sites are good entry points for finding good addresses on the Web. Here it is more feeling and luck which will enable you to uncover rare sites or unheard-of services. So, if you find the rigorous methods of the large search engines tiresome, allow yourself to be guided and to ride the wave.

▬▬▬ Cool Site of the day

http://cool.infi.net/

This selection, made by Glenn Davis of Infinet, is without doubt one of the best starting points on the Web. Each day you can discover a new "Cool Site".

Figure 9.1: The Cool Site of the Day home page

▬▬▬ Magellan – Search Voyeur

http://www.mckinley.com/

With Magellan's Search Voyeur module 12 themes are offered at random, with the offer being refreshed every 15 seconds. Choose a theme and click on the Search button. Magellan will then send you the corresponding sites in order of relevance. Ideal for surfing without any preconceived ideas!

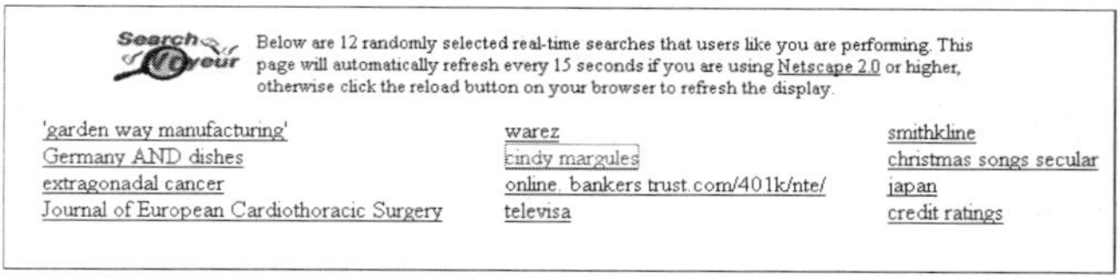
Search Voyeur
Below are 12 randomly selected real-time searches that users like you are performing. This page will automatically refresh every 15 seconds if you are using Netscape 2.0 or higher, otherwise click the reload button on your browser to refresh the display.
'garden way manufacturing'
Germany AND dishes
extragonadal cancer
Journal of European Cardiothoracic Surgery
warez
cindy margules
online. bankers trust.com/401k/nte/
televisa
smithkline
christmas songs secular
japan
credit ratings

Figure 9.2: Let yourself be guided by Magellan!

Discovering Web site compilations and guides

Guides offer rational surfing. They generally include access by subject and search tools. On the Web, everybody, or nearly everybody, draws up lists of sites to visit and all this may more or less resemble a guide. You will find lists of good addresses on the home page of other surfers. In the light of this profusion of guides, which we have had the opportunity to present during previous chapters, we have decided to be restrained (for once!) and provide you with just one good address.

MSN Directory

http://msn.co.uk

The MSN Directory, published by Microsoft, lists hundreds of top sites for UK surfers in over 100 categories. To browse the directory after arriving at this page, either choose one of the 13 major categories from the pull-down list at the left of the page, or choose the Index Page and click the "Go" button to see all of those more specific categories.

In many of the categories you'll find not only links to top web sites, but direct links to the top news stories in the national or computing press, reviews of the latest computing products, special offers from UK shopping sites, and many other time-saving links. Many of the category pages include simple search forms that let you search particular sites without leaving the MSN directory.

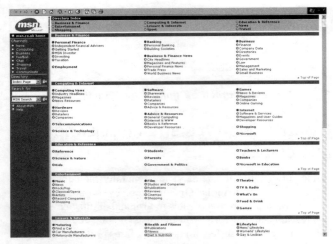

Figure 9.3: Hundreds of ready-sorted links and searches at the MSN Directory

Compilations to visit

- *Top 50 UK Web Sites: **http://www.top50.co.uk***

- *PC Magazine's Top 100: **http://www.zdnet.com/pcmag/ special/web100***

- *The Weekly Hot 100: **http://www.100hot.com***

VISITING TOP SITES

On the Web you will find all kinds of Web site classification, such as:

- classifications carried out by surfers who choose the best sites;

- classifications prepared by professional teams (cyber-journalists);

- classifications prepared by a software based on how often a site is visited, generally known as hits.

Here is one example:

▬▬▬ Lycos Top 5%

http://www.pointcom.com/

The best of the Web by the Lycos team.

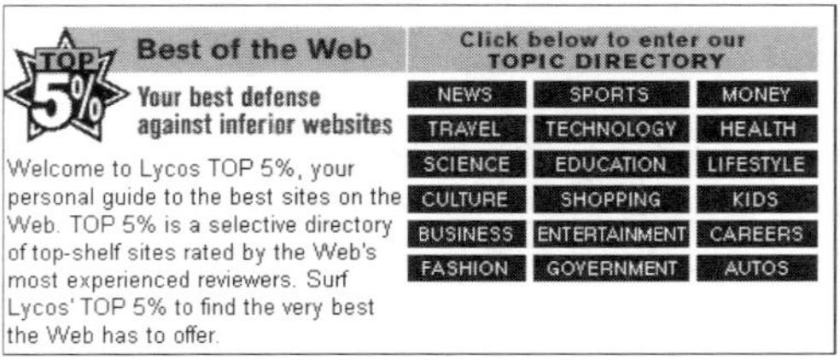

Figure 9.5: Lycos' Top 5%

BROWSING THE WHOLE WORLD

You have had the opportunity to travel the world using the large international search engines. If you want to go further still in your trip into cyberspace, there is an impressive number of specialist tools which will take you from one country to another with the greatest of ease. A world tour in the space of a few clicks!

▬▬▬ Matilda

http://www.aaa.com.au/images/logos/searches/world/

This travelling search engine comes to us from Australia. With more than 200 countries listed and endowed with specific search mechanisms for each of them, Matilda is worth discovering.

You will be astonished to note that lists of hits provide very specific information, including the target audience and even the site's popularity!

Figure 9.6: The choice of countries offered by Matilda

Figure 9.7: Searching by country; in this case, Benin

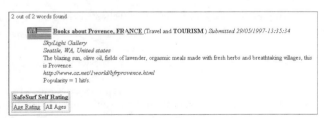

Figure 9.8: An answer from Matilda

Woyaa!

Http://www.woyaa.com

Figure 9.9: The home page of the African search engine Woyaa!

Woyaa! Is a specialist search engine for Africa, with subject entries (arts, society, culture, etc.) or keyword searching. A considerable number of sites are listed, and if you want to carry out research into the African continent, this is where to start.

Searches are made by subject or keyword.

As a bonus, the site offers you Net Radio Earthbeat so you can surf to music!

Figure 9.10: The result of a search on the keyword "Sierra Leone"

The Virtual Tourist

http://www.vtourist.com/webmap/

Click on the globe and then on the map of the region of the world selected. You then access specific resources for the country chosen, in English or in the language of the country in question.

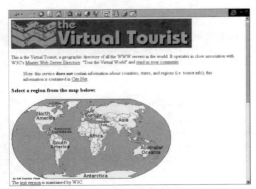

Figure 9.11: Click on the world map to access a continent or a country

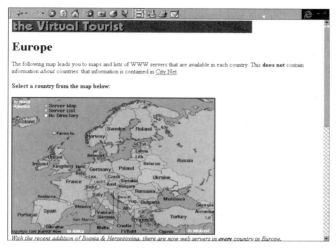

Figure 9.12: Access to Europe

Select a country from the map below:

Figure 9.13: Here, access to the countries of South America

Country Maps for Europe

http://www.tur.nl/europe/

Developed in the Netherlands in English, this site offers a list of Web sites for all the countries of Europe. You select from a map or by choosing the name of the country from a list. The site offers other search tools in order for you to discover the country.

Figure 9.14: Selecting a country

SEARCHING IN ANY LANGUAGE

What we offer here is a selection of languages and sites. A quick tour of the Web to allow you to glimpse its wealth. Also, you must set the parameters of your browser for it to correctly display the characters of the various languages.

In all languages

EuroSeek

http://www.euroseek.net/page?ifl=uk

This is a remarkable search tool which allows you to target a search
on a given country and also to choose the answers (target sites) in
the language of your choice. 40 languages are offered, classified in
alphabetical order from Bulgarian (Balgarski) to Turkish (Türkçe)!

*Figure 9.15: Searching throughout all the countries and
in all the languages of Europe*

Figure 9.16: Searching sites in Russia

In English, for non-English speaking countries

Russia on the net

http://www.ru/

The home page of Russia on the net is shown in Figure 9.16.

Jewish Communication Network

http://www.jcn18.com/scripts/jcn18/paper/query.asp

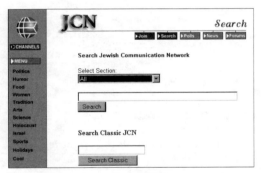

Figure 9.17: A search tool for the Jewish community

Indonesia Net

http://www.indonesianet.com/search.htm

Figure 9.18: Searching in Indonesia

▰▰▰ In Austrian

Austrian Home Page Search Engine

http://www.aco.net/Server-in-AT/
Plain text searching.

Austronaut

http://austronaut.at/url.ims
Searching by URL or keyword.

Figure 9.19: Searching for Austrian sites

▰▰▰ In Danish

Jubii

http://www.jubii.dk
Searching by keyword and by subject.

Figure 9.20: Searching in Danish

In Spanish

University of Cordoba

http://www.uco.es/

In Portuguese

University of Aveiro

http://www.ua.pt/

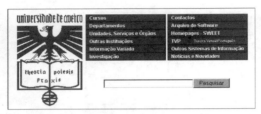

Figure 9.21: Searching in Portuguese

Chapter 10

Searching in the media

THE CONTENTS FOR THIS CHAPTER

- Getting back to the source with press agencies
- Rummaging through newspapers and magazines
- Searching the international media
- Viewing television broadcasts on the Web
- Listening to live radio throughout the world

Searching in the media is a highly efficient way of obtaining information on the Web. Here "media" is taken to mean all the traditional large-scale media (press, radio, television) with Web links, but also cybermedia which have developed on the Web.

Having made this distinction, it must be said that the traditional media have managed to assert themselves on the Web, while many cybermedia have yet to prove themselves.

Online media thus offer the latest news items and also open up their archives. A new track for finding information!

GETTING BACK TO THE SOURCE WITH PRESS AGENCIES

If you are interested in current affairs, start here. You can obtain information well before the 9 o'clock television news. And above all, you can find dispatches from all over the world. On the Internet, you choose what you want to receive, and you are not governed by the choices made by professionals in the world of communications. But beware, some services are, quite legitimately, subject to a fee.

PA News Centre

http://www.pa.press.net

This UK Press Association web site should be your first port of call to find the latest national, international and business news stories. A "StoryFinder" service allows you to run a keyword search for news articles, optionally selecting an area of the UK or a particular topic to narrow the search.

Reuters

http://www.reuters.com

This is one of the largest press agencies in the world, and the service offered by the Web is high quality. The options for searching by subject are particularly attractive. For this you use the drop-down list above the Products button.

Figure 10.1: Catch the breaking news stories first at the PA News Centre

Figure 10.2: The Reuters agency opens its doors to you

Yes, information is a product! But beware of small configurations. Multi-windowing and animations can slow down your consultation of the service.

Figure 10.3: Searching for sports news items

Computer Wire

http://www.computerwire.com/

Figure 10.4: Searching through the archives with computer Wire

Computer Wire is a press agency specialising in information techniques. The site even allows you to create your personalised information letter. The HyperSearch tool allows you to search using several criteria in the archives of 10 magazines devoted to computers and telecommunications.

Finding press agencies with Yahoo!
*To go even further, you can use Yahoo! to ask for a list of the world's press agencies. The address is: **http:// www.yahoo.com/News_and_Media/Newswires/ index.html.***

RUMMAGING THROUGH NEWSPAPERS AND MAGAZINES

If you want to find a news item or article, then consult the Web sites of the large dailies and explore their archives when they are available online. You can even set up electronic press reviews. You need to identify good sources and learn to rummage around, searching by criteria or in plain text mode (technique of indexing all significant words contained in an article). Head for the archives!

The Electronic Telegraph

http://www.telegraph.co.uk

The online edition of the popular Daily Telegraph allows searching of its news archives dating back to 1994, along with separate databases covering travel reports and book reviews. To use this service, you will first need to register with the site. Registration is quick and free, and simply involves providing some personal details. You can then choose a personal user name and password which provide access to the archives.

The search options allow you to:

- search for names (using initial capitals) or keywords;

- specify a particular year, month or date; or

- narrow your search to a single section.

When searching by section, you can hold down the Ctrl key to select multiple sections for searching.

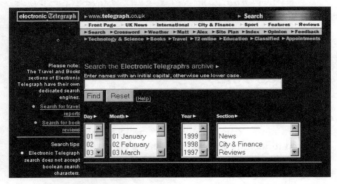

Figure 10.5: Searching for articles in the Electronic Telegraph

Results are presented in a list comprising:

- the date;

- the headline of the article; and

- a relevance score in percent.

Just click on the headline to open the article.

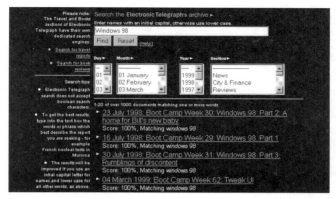

Figure 10.6: The results of an Electronic Telegraph searching using the keywords "Windows 98"

The Times

http://www.the-times.co.uk

If you omit the hyphen, you'll arrive at the home page of News International, providing links to The Sunday Times, The Times Literary Supplement, The Sun, the News of the World and other publication. Although The Times itself offers no keyword search, its Library section allows you to browse back issues from 1996, and a useful InfoTimes section gathers a number of past articles together into categories for research purposes.

The Guardian/The Observer

http://www.guardian.co.uk

Although its archives date back only to September 1998, the Guardian's search options are more powerful than those of some of its rivals. Searches automatically cover both Guardian and Observer archives in the period you select, and results can be sorted by date or by relevance to your keywords. You can also choose to display only the first paragraph of matching articles, making it easy to see whether you've found what you were looking for.

Figure 10.7: Search Guardian and Observer archives using Boolean and case-sensitive keywords

▬▬▬ Financial Times

http://www.ft.com

Like the Electronic Telegraph, mentioned earlier, access to the FT site is free but you must first "subscribe" by filling in a few personal details and choosing a password. Once you've logged in to the site, enter the Archive section to search over 4 million articles from a number of financial sources dating back to 1996. Among the many search options available, you can:

- choose whether to search in articles' headlines, text or both;

- choose the types of publication search or pick a single publication (such as the FT itself) from a list of over 3000;

- search a particular financial sector (such as Banking or Media);

- search by region (such as UK & Ireland or North America).

Search by publication   new saved searches account tips

Search over 4m articles from 3,000 publications such as
Independent on Sunday - United Kingdom, Saudi Gazette - Middle East Newsfile
and The Sunday Telegraph (United Kingdom)

Type the words you wish to search for in the box below.
[   ] Search

Find articles that contain...   In the article's...
 ⊙ all words entered above   ⊙ headline and text
 ○ at least one of the above   ○ headline only
 ○ exact phrase entered above   ○ text only

Published in: [Last 3 months ▾]   Sort by: [Date ▾]

Search by publication   or   Search by sector & region
All Clear

 ☑ FT Newspaper List   ☑ Global Business List
 ☑ Top Sources Worldwide List   ☑ US Business List
 ☑ News Wires List   ☑ European Business List
 ☑ Press Releases List   ☑ Asian Business List

You will search on [3354] publications

[Save search]   [Search]

Figure 10.8: The world's financial news at your fingertips, courtesy of the Financial Times

SEARCHING THE INTERNATIONAL MEDIA

There are thousands of titles to discover on the Web. It is not easy to find your way, but information is there somewhere. Here too we will restrict ourselves to a selection of the main titles on the Web (such as The *Washington Post*), or new media (such as msnBC). While the traditional media is sometimes reproached for giving disinformation, one could accuse the Web of over-information. For this reason the search tools and filters which it allows you to use are useful. That is also why it may be useful to target a few good addresses to serve as a starting point for looking for specific information.

▬▬▬ msnBC

http://www.msnbc.com/

Born out of collaboration between ABC and msn (Microsoft Network), as soon as you reach the site msnBC offers to install a news browser. This plug-in for your browser will be of use to you in looking at international current affairs.

Beware; you will need a good configuration in order to benefit fully from this very multimedia site.

Figure 10.9: Welcome to msnBC

The Washington Post

http://www.washingtonpost.com/

When you have the front page online, you can:

- select a section from the drop-down list;
- activate the paper's index;
- start the search engine.

It is best to search for articles over the last fortnight, but you can restrict this time band.

The search page also gives access to:

- the news dispatches of the AP agency (Associated Press);
- a multitude of databases in the business, leisure and entertainment domain, e.g. Video Finder gives you access to 3,600 films on video with a report on each one.

You can also search for news by country. 220 countries are listed and if you type "France", for example, you will access a front page relating to France which is particularly rich. You can also travel the entire world!

Figure 10.10: The Washington Post home page

Find Washington Post Articles

Use this form to find Washington Post stories from the **past two weeks**. Searches and story retrieval on this page are free.

Advanced Search Tips

Headline:	
Author:	
Date Range:	Within 14 Days ▾
Any Word(s) in the Article:	

TIP: To find words used together in an article, you need to put single quote marks around the entire phrase.

[Search] [Reset]

Figure 10.11: Searching for articles in the Washington Post

Figure 10.12: Searching for a film by title

▬▬ The Chicago Tribune

http://www.chicago.tribune.com/

The Chicago Tribune is another heavyweight of the media with sophisticated search tools. The online version allows you to search its archives going back to 1985.

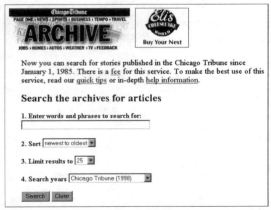

Figure 10.13: Searching for articles in the Chicago Tribune

You can ask for the following results to be sorted:

- date;
- relevance;
- frequency of appearance of the keyword.

You can search in relation to a specific year, or search through all the archives. It is possible in this way to obtain nearly 3,000 published articles which included the word "Microsoft". For each article you get a summary. You will have to enter a command in order to obtain the full version.

An international videotex service on your PC!
*Webdo, a Swiss magazine, offers an impressive list of online media **(http://www.webdo.ch/webactu/ webactu_presse.html)**. You search by country and have access to thousands of titles!*

*Also take a close look at Chaplin's News **(http:// www.geocities.com/ Heartland/2308/)**. The list of subjects is impressive.*

*Use Net Media **(http://www.go-public.com/netmedia/)** if you want to search by country.*

*Or again, with classification by country, The Inkternational News Link **(http://inkpot.com/news/)**.*

Television and radio on the Web
If you are looking for information of a multimedia nature, plug into online television and radio. The former offer extracts from their broadcasts, while the latter can offer live broadcasting. Technology allows this and Internet Explorer 4.0 has the Progressive Networks' Real Player which is necessary in order to listen to radio on the Web, as a standard feature.

To follow: CNN (http://www.cnn.com/), NBC (http://www.ultimatetv.com/tv/us/networks/nbc.html) and the large UK channels which all have an active presence on the Web.

If you like to "channel surf", plug into Ultimate TV Webcasting (http://www.ultimatetv.com/webcasting/events/streaming_v_intl.ht). The site offers you links with television channels around the world, with live video for each of them. Version 4 of Real Player also allows you to view videos.

Chapter 11

"Push" technologies

THE CONTENTS FOR THIS CHAPTER

- Using personalised newspapers
- Receiving information by e-mail
- Finding mailing lists
- Discovering helper applications
- Getting to grips with intelligent agents

While we have learned to "pull" information from the Web during the first ten chapters of this book, the last two chapters will be devoted to push techniques. This time we do not go to the information, the information comes to us; it is "pushed" onto our PC.

While Microsoft introduced the concept of CDF channels with Internet Explorer 4.0 (see next chapter), many other tools allow us to retrieve information automatically.

The following examples should be noted:

- the possibility of defining one's preferences and receiving personalised newspapers;

- information delivery services by e-mail;

- helper applications and other intelligent agents which ferret through the Web for you.

With "push" technologies, the Web is trying to match the media (which broadcast information), but the major difference is that in this instance the information may be personalised. This is one of the key aspects of services and media on the Internet.

USING PERSONALISED NEWSPAPERS

Consulting a newspaper whose front page and articles are intended for you personally is not just a dream. On the Web there are many services which allow you to define your tastes and preferences. Knowing your profile, these services will then be able to deliver truly personalised information.

▬▬ Crayon

http://crayon.net/

With Crayon, step by step you are going to construct your own live daily newspaper free of charge. A password, and then you proceed with the choice of a title, formatting, headings, links with information sources, etc. It just takes a few minutes, so discover Crayon without delay!

Two design modes are offered:

- normal mode, which scans all information sources in detail;
- quick mode, for the basics (it uses the American post code to deliver regional information).

By preference use normal mode and select:

- headings (world, tech, science, religion, Web, etc.) by clicking on the colour crayons at the foot of the page;
- information sources included under a heading by checking the media which interest you.

You can also add your suggested information sources. Finally, click on the "Create my own newspaper now" button when your are ready!

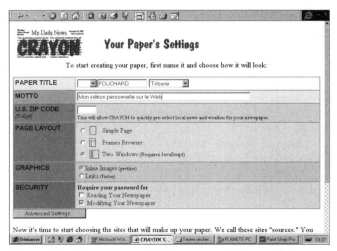

Figure 11.1: Crayon is useful for personalising your newspaper

Then select the hierarchy of your headings (the different sections about which you have already given information). Click on "Publish my newspaper" and it is done. The paper can be consulted immediately. Do not forget to save the page in your favourites in order to access it more easily in the future.

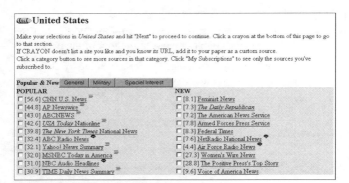

Figure 11.2: The interactive design of a personalised newspaper

Figure 11.3: The front page of a personal newspaper

▬▬ MyYahoo

http://edit.my.yahoo.com/config/login

The famous search engine also offers an à la carte information service. Here too you are going to create your own personalised paper incorporating, for example:

- stock exchange information;
- sports results;
- international current affairs;
- Web resources;
- and much more!

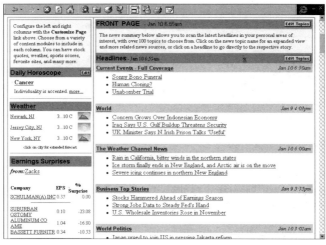

Figure 11.4: An example of a personalised paper using MyYahoo

Follow the step by step instructions. From weather reports to computers via sports or television, the list of headings is enormous.

Then, MyYahoo takes care of the content and you just have to plug into your daily newspaper. If you are permanently connected with the Internet (via cable), you can receive information on your desk in real time. Otherwise, you can download software, News Ticker, which will enable you to access the information with just one click.

Excite

http//www.excite.com/Info/

Yahoo! is not the only search engine which tracks down information for you. Excite offers a similar service with 14 subject channels at your disposal. After identifying yourself, you choose your channels. The Web pages offered can be customised to your requirements.

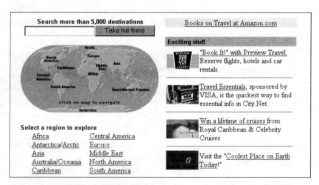

Figure 11.5: Searching Excite's Travel channel - 5,000 possible destinations!

News Tracker

http://nt.excite.com/

This is a new service with the Excite hallmark, which allows you to keep track of current affairs. This gigantic information service compiles more than 300 sources. All current affairs are covered, including computer news. As with a search engine, you can search for information by indicating a few keywords. You can create your "personalised newspaper" by defining your areas of interest.

Figure 11.6: An extract from the News Tracker search page. The data entry zone is right at the bottom of the page!

NewsHub

http://www.newshub.com/tech/

This is not, strictly speaking, a personalised newspaper. NewsHub will track down the titles of the largest information sources in the Hi-tech domain every 15 minutes. If you want a lower frequency, you can indicate this under Preferences.

Thus, NewsHub gathers only computer news by scanning information that you can find on sites such as Yahoo's Tech, Nando InfoTech, Tec Wire, Gina I-Wire, InfoWorld, PC Week, ZD-net, ComputerWorld or Wired, to name but a few.

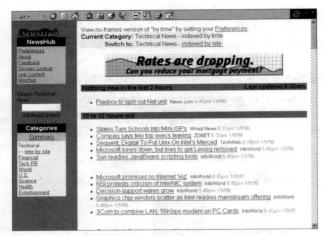

Figure 11.7: NewsHub tracks down information for you

RECEIVING INFORMATION BY E-MAIL

A very easy way of tracking down information is to subscribe to e-mail delivery services, of which there are many on the Web. Use them, as they are often very useful. But do not overuse them or your mail box will be snowed under. You can always cancel a subscription by sending a command along the lines of "unsubscribe" to the server.

InBox Direct

http://form.netscape.com/ibd/html/ibd_frameset.html

This service offered by Netscape, the publisher of the browser with the same name, allows you to receive information directly to your e-mail in-box. You just have to check the publications, classified by heading, which interest you. They will then be delivered to you by e-mail in HTML format. Just check that your messaging software accepts this format (the latest generation software does accept this format, e.g. Netscape Communicator and Internet Explorer 4.0 or 5.0).

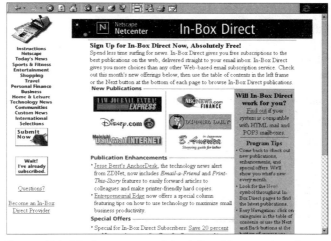

Figure 11.8: The home page of the InBox Direct messaging service

Figure 11.9: Check the publications and services of your choice; in this case, Travel

Slate

http://www.slate.com

This magazine, which belongs to Microsoft, was not designed to relay computer news. It is a general magazine which covers politics as well as culture or leisure.

You can use Slate in three ways:

- just consult the Web site;

- leave your e-mail and you will receive the titles of the articles when they appear. This enables you to go to the site only when you have found interesting articles;

- ask to automatically receive the full edition of Slate in your mail box. You will then receive a file in Word format which you can consult on screen or print.

Figure 11.10: Use the Services button to discover the possibilities of the Slate magazine

International mailing lists
To find international mailing lists, plug into the Listz site
(http://www.liszt.com/). It lists more than 80,000 mailing
lists. You can search by keyword or by subject.

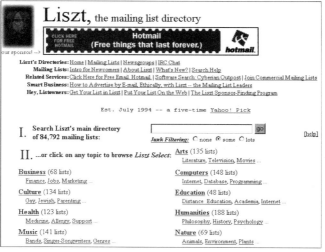

Figure 11.11: Searching for worldwide mailing lists

DISCOVERING HELPER APPLICATIONS

Here we list all the other push technologies which we feel are of
interest, except for CDF channels which we will deal with in the
final chapter. We should point out that Netscape, Microsoft's main
competitor, also has its own push technology: NetCaster.

Known as helper applications or intelligent agents, they are at
everybody's beck and call and are the future of the Internet. Learn
to use them and to know their strong points and weaknesses.

PointCast

http://www.pointcast.com/

This is one of the Web's leading helper applications. It allows you to explore the Web's sites in search of information which interests you. To do this you will have to download the software from the site. You must then set the parameters according to your tastes and requirements. Then, just start PointCast to obtain the information which interests you. Better still, you can also set the program's parameters so that it starts automatically. You must also tell it how often or when to update itself.

BackWeb

http://www.backweb.com/

Designed by an Israeli company, BackWeb brings you information on a plate. Each information provider in effect creates a channel through which it is able to "push" its information, a little like a television channel. You thus receive the selected information directly onto your PC. In addition, this transmission is optimised, since information is only sent when the connection is not in use. Also noteworthy is the fact that you will be notified of the arrival of any new information. It is up to you to choose the channels from among the different headings available.

Intelligent agents
These are software – free or otherwise – which you must download and which will then carry out searches on the Web for you.

They include Autonomy (http://www.agentware.com/) which allows you to "train" specialist agents for a task. These agents then undertake to find the information you want. It is quite a long process and is not suited to occasional searching. Train an agent for a clearly defined task and it will be of use for several months.

It is also worth getting to grips with Wise Wire (http:// www.wisewire.com/), Bot Spot (http://www.botspot.com/), Copernic (http://www.copernic.com/) or Agent Intelligent (http://www.pls.com/products/agent/). It is up to you to find the right one to do the job.

Also discover Teleport Pro (http://www.tenmax.com/ pro.html), a tireless information hunter which can download entire Web sites onto your PC! Use with care!

Figure 11.13: The choice of an information channel on BackWeb

Chapter 12

Active channels

THE CONTENTS FOR THIS CHAPTER

- The concept of Web channels (Active Channels)
- Subscribing to a channel
- Using an online guide, Channel Guide
- Worldwide channels

The concept of Web Channels (*Active Channels*), developed by Microsoft in Internet Explorer 4.0, is an application of what are known as "push" technologies (as opposed to "pull" technologies). Either you go to information and pump your favourite search engine, or information will come to you (with your consent), without you needing to do anything (other than choose your information channels and plan the arrival of information on your PC).

Searching the Internet

Subscribing and downloading sites are similar, underlying notions. From Internet Explorer 4.0 you can naturally subscribe to the contents offered, but in the same way you can also download Web sites to meet your requirements.

The differences between a Web channel and any other Web site to which you can subscribe are as follows:

- **Technological difference**. Web channels conform to a development standard developed by Microsoft (CDF: *Channel Definition Format*).

- **Marketing difference**. Web channels use an immediately identifiable subscription button. Some of them even have privileged agreements with Microsoft, so as to appear directly on your Desktop after you have installed Internet Explorer 4.0.

Figure 12.1: The "remote control" of channels previewed on the Windows desktop

INTERACTIVE WEB CHANNELS

The main novelty may be considered to be the bringing together of medium and media, container and content, software tool and Web sites. These contents are the result of partnerships (access displayed on the Windows Desktop). In total there are 10 active channels which are presented on the Windows Desktop.

And, as the development of channels is not limited to the initial introductory agreements, which gave preference to UK contents, we must add the Web channels of the whole planet.

How to subscribe

Figure 12.2: The home page of the BBC service

It is relatively simple to subscribe to a channel accessible on the Desktop, even if the initial mechanism involves a considerable number of stages. First of all, click on one of the buttons shown. In our example we have chosen to subscribe to the BBC channel.

 Searching the Internet

One click on the BBC button and the Internet connection is set up. At the same time, the home page opens in the browser. There is a blue button on this page: Add Active channel. This allows you to start all the procedures.

Figure 12.3: The Web channel subscription button

Since you have decided to subscribe to a new channel, the updating procedure is initiated. Immediately afterwards the subscription management window opens.

Figure 12.4: Channels in the process of updating

Figure 12.5: The subscription procedure

You are offered three choices:

- Only when I choose Synchronise from the Tools menu;
- I would like to create a new schedule;
- Using the existing schedule.

If you opt for the second choice you can then choose the frequency and downloading schedule yourself. Choose the first option to download the updated contents manually whenever you like, or the third to use the downloading schedule recommended by the channel publisher.

All planning data will be gathered by the subscription wizard. You can go over the entire procedure in greater detail by referring to Chapter 7.

Figure 12.6: Starting the subscription wizard

At the end of the planning phase, the software will possibly ask you if you wish to update your screensaver.

Screensaver
Your screensaver is produced by setting the display parameters so that a dynamic page is displayed on screen after a certain length of time without any keystrokes or mouse manipulations. You stipulate this time yourself. Thus,

when your PC is inactive, it displays the selected screensaver, in this case a presentation of the Web channel to which you have just subscribed. It is a good way of encouraging you to interface regularly.

Figure 12.7: Updating the screensaver

Once this adjustment of the screensaver is validated, the channel to which you have just subscribed is finally displayed.

*In order to adjust the screensaver, right-click on a free area of the Windows Desktop. Select **Properties** and click on the Screensaver tab. Set the parameters: display after so many minutes of inactivity, etc.*

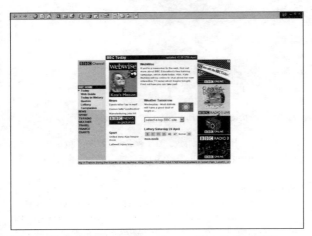

Figure 12.8: Opening of the channel, here, the BBC

Index

 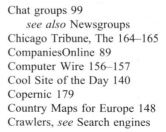

Searching the Internet

Searching the Internet

 Searching the Internet

 Searching the Internet

 Searching the Internet

 Searching the Internet

 Searching the Internet

 T

Teleport Pro 179
Television and radio 165–166
Times, The 159

 U

UK
  newspapers 157–160
  search engines 30–43, 78–80
UK Index 41–42
UK Plus 23, 38–39
UK Shopping City 137
UK Street Map Page 121–122
University of Aveiro 152
University of Cordoba 152
URL 15

V

Virtual libraries 124–128
Virtual shops
  books 130–136
  general shops 137–138
  search engines 138
Virtual Tourist, The 146–147

 W

Washington Post, The 162–164

Waterstone's Online 131–132
Web address structure
  anchors 16
  http:// 15
  www 16
WebCrawler 12–13, 57–58, 62, 138
Webseek 111–113
Web Taxi 75
WhoWhere 82–83
Wildcards, see Searching
Windows 95
  Importing images 110
Wise Wire 179
Worldwide search engines 45–61
World Wide Web Yellow Pages 93
Woyaa! 145–146

 X

Xerox Map Viewer 118

 Y

Yahoo! 23, 50–51, 61
  Image Surfer 114–115
  My Yahoo 171–172
  searching for other search engines 75–76
  UK and Ireland 30–32
Yell 23, 39–40, 90–91
Yellow pages 90–95

 192